Rafael Novoa

Agronomic principles: basis for agronomic theory

Rafael Novoa

Agronomic principles: basis for agronomic theory

ScienciaScripts

Imprint
Any brand names and product names mentioned in this book are subject to trademark, brand or patent protection and are trademarks or registered trademarks of their respective holders. The use of brand names, product names, common names, trade names, product descriptions etc. even without a particular marking in this work is in no way to be construed to mean that such names may be regarded as unrestricted in respect of trademark and brand protection legislation and could thus be used by anyone.

Cover image: www.ingimage.com

This book is a translation from the original published under ISBN 978-620-0-05638-2.

Publisher:
Sciencia Scripts
is a trademark of
Dodo Books Indian Ocean Ltd. and OmniScriptum S.R.L publishing group

120 High Road, East Finchley, London, N2 9ED, United Kingdom
Str. Armeneasca 28/1, office 1, Chisinau MD-2012, Republic of Moldova, Europe
Printed at: see last page
ISBN: 978-620-7-39191-2

TABLE OF CONTENTS

Summary

An agronomic theory summarising the knowledge acquired to date is of great theoretical and practical importance. The progress of the sciences from their infancy to their maturity follows a path that leads them from a descriptive language of principles to a more concise and precise mathematical language of laws to finally arrive at models that formally integrate the knowledge that we can basically call a theory. Thus, the three basic agronomic principles discussed are: the use of efficient improved organisms, giving these organisms the right ecological conditions for the expression of their genetic potential, and appropriate management.
It appears that the influence on the yields of each of these principles is on average about 33 % each. The practical application of these principles is illustrated for three cases: the effect of agricultural technologies on the environment, how they can help to analyse the management of a crop and to visualise the relationship between technologies, products and the agronomic sciences behind them. Finally, the agronomic laws are discussed, and finally the simulation models, their types, and the dynamic simulation models are validated, and their advantages, problems and uses are discussed.

Keywords: principles, agronomy, agronomic laws, dynamic models.

Summary

An agronomic theory that sum up the acquired knowledge up to date is of great theoretical and practical importance. The advance of sciences from its beginnings to it maturity seems to follow a path that star with a descriptive language of principles to a mathematical language more concise and precise to finally arrive to models that formally integrate all the knowledge into a theory. So, the basic agronomic principles are: use efficient improved organisms, provide to these organisms the ecological conditions that are appropriate to let them express their genetic potential and provide a good management.
It seems that influence on yield of each of these principles is for a 33 % each. The practical use of these principles is illustrated for three cases: the environmental impact of the agricultural technologies, an analysis of the way a crop is managed and the relationship of agricultural technologies, products and agricultural sciences that sustain them.
Agronomic laws and simulation models are discussed. The kind of models, the way dynamic models are build, how they are they validated, their advantage and problems, and finally the use of models are discussed.

Chapter 1

1. - INTRODUCTION.

The main objective of this work is to integrate the advances obtained in the agronomic field, oriented more towards an understanding of the core processes than towards a detailed knowledge which is left to the specialist in one of the agronomic sciences. The aim is to lay the foundations of an agronomic theory. Obviously this is an initial step - a starting point which will undoubtedly be further developed in the future.

The plan is as follows: describe the basic principles and their laws, give examples of practical uses of these principles, discuss some advances in quantifying the impact of these principles on production, review the significance of the use of simulation models. The term agronomy (derived from the Latin combination ager, field y from the Greek, nomos, law) will be used in its broadest sense which includes not only the cultivation of land to generate plant-derived goods but also to generate animal-derived products. However, the examples I will cite are all from the plant area, but this does not mean that these principles are not valid in the area of animal production. In fact they are valid for any productive activity that uses living organisms to produce goods.

1.1. - Science, Theories y Agronomy.

The success of various sciences such as physics or genetics, the basis of spectacular technological developments, is closely related to the fact of having solid theories. These represent a summary of the advances made by experimental results and the inferences or deductions of researchers, made to date, in a given science. Constructing theories is a game of data compression. It is finding a message, as short as possible, which when unpacked produces an accurate model сото explains Wilczek, 2008. A good scientific theory enables technological breakthroughs that follow from its consequences and is therefore of enormous practical utility. A notable example of this is the discovery of the structure of DNA, which has been the basis for practical developments in numerous fields of biology. Moreover, by following its principles and laws, researchers can advance on a solid basis in the development of that science. If agronomy is compared with breeding, which is based on the application of genetic laws and is currently in its infancy in the application of genetic engineering, it can be seen that its advances are much more predictable and safer than those in agronomy. This is attributable to the fact that genetics has developed a theory and laws which, when applied, lead, with complete certainty, to the production of "improved" plants and animals, i.e. organisms with consciously introduced characteristics which give them the potential to produce more and better quality goods. In agronomy, there is no guide similar to genetic theory, there is no explicitly recognised agronomic theory.

Perhaps this is why the Department of Agronomy at the University of California, Davis, states that Agronomy is the science and **art** of crop production. It is possible that in this definition it is recognised that there is art probably for lack of a good scientific theory.

Sciences are usually classified as natural (e.g. physics, chemistry, biology), social (e.g. economics, sociology) and formal (e.g. mathematics, computer science). According to the above, agronomy y would be placed within the natural sciences.

One might ask whether agronomy is a science or not. Science, *scientia* in Latin, meant only knowledge until the Middle Ages. Subsequently it came to mean, in addition, a precise and standardised way of obtaining knowledge. This means to submit our hypotheses to corroboration by experiment, i.e. to the use of the scientific method. However, not all sciences use experimentation.

Thus, mathematics uses a logico-deductive methodology and astronomy resorts to observations to confirm its hypotheses.

If we understand science as a body of knowledge acquired using the scientific method, then agronomy is a science because its knowledge is only accepted if it is obtained by following this procedure and can be verified by third parties.

If you look at physics, perhaps the most advanced science today, you can see that in addition to following the scientific method it has established principles, such as uncertainty, has used the language of mathematics to express its laws, and has recognised that it is possible to develop theories by reasoning alone (theoretical physics) provided that they agree with known experimental results. Likewise, biology has been actively exploring the field of theoretical biology, as evidenced by the existence of the journal como the Journal of Theoretical Biology. There are also departments and centres of Theoretical Ecology at the universities of Wageningen, Netherlands, Lunds, Sweden, Helsinki, Finland, Leuven, Belgium, Queensland, Australia, Princeton, Stanford Universities, USA and many others. In general these groups work with mathematical models to summarise, quantify and understand biological problems. System Ecology" departments have also been formed that apply general systems theory and modelling to ecology como those of the University of Stockholm o of Bucharest o the Vrieje University of Amsterdam o system ecology programmes como that of the University of Florida in the USA.

Ecology is interesting because it is a biological science very close to agronomy, since it focuses its work on the interactions between populations, not individuals, and the surrounding environment, including populations of other plants, micro-organisms, diseases and pests. It has been said that agronomy is ecology plus economics, but in reality it is rather autoecology (which has the advantage of allowing experiments to test our agronomic hypotheses) and more than economics it is management (which includes economics, sociology and other relevant disciplines).

Therefore we could define agronomy as como: The Science of Agricultural Ecosystems and their Management. These ecosystems are those formed, mainly, by populations of one or a few species, oriented to the production of goods of human interest (food, energy, fibres, wood, drugs, pigments and others).

It is also a science that integrates knowledge generated by many disciplines: physics, chemistry, biology, physiology, biochemistry, genetics, meteorology, soil science, pathology, bacteriology, nutrition, entomology, ecology, economics and geography, among others. This integration often takes information at the molecular, tissue or organism level and gives it meaning at the population or agricultural level.

1.2. - Agriculture y Agronomy

On the other hand, agriculture, the field of application of agronomy, can be considered como a form of symbiosis occurring in at least four groups of animal animate species: ants, termites, bark insects of the subfamily Scolitinae, and, by man, Shulz & Brady, 2008. Although there is recent evidence of human consumption of uncultivated grains processed around 30000 years ago (Revedina et al., 2010), man-made agriculture appeared only around 10,000 years ago (Zeder, 2008, Miller, 2008, Barton et al., 2009) but it was born much earlier, around 50 million years ago in the case of some ants. Numerous species (about 160 in the case of ants of the Attini tribe alone (Shulz & Brady, 2008, Benckiser, 2010) practice it with varying degrees of sophistication not only by cultivating fungi or yeasts but also by doing a kind of cattle farming como lo is the case with ants and aphids. There are various types of agriculture made by insects to produce food. In the case of ants alone, at least five types are recognised: the cultivation of fungi that retain their ability to live independently, apparently the most primitive; the cultivation of coral fungi (of the family Pterulaceae); the cultivation of yeasts; the cultivation of "domesticated" or "selected" fungi unable to live independently but capable of producing swollen hyphae that the ants harvest as food; and that of leaf harvesting, of more recent appearance, 8 to 12 million years ago. There is also an ant-aphid symbiosis in which the ants provide protection for the aphids by taking advantage of the sugar secretion they produce, which can be classed as a type of insect farming. Non-human agriculture necessarily implies a certain, apparently rudimentary, agronomic knowledge, being a highly organised and sustainable system (Benckiser, 2010). Modern man-made agriculture, unlike that of other species, is guided by scientific agronomic knowledge and is estimated to have started about 10,000 years ago (Zeder, 2008, Miller, 2008. Price 2009, Barton 2009).

However, given that agriculture is only done by animals that are organised in societies, it can also be said to have a social basis y in this sense agronomy shares a place with economics.

1.3. - The scientific method applied to agronomy.

As in any science, agronomy uses the so-called scientific method to establish how good its hypotheses are. These are tested in specially designed experiments which must have the necessary treatments, controls (treatments applied at random, a minimum of three replicates and controls), measurements and statistical analysis to avoid making errors or giving true results due to chance alone. The latter analysis applies the null hypothesis that any hypothesis is true until some statistical evidence (derived from experiments or otherwise) shows otherwise. Although the truth of a hypothesis cannot be established on the basis of experiment when the null hypothesis is true, the hypothesis is provisionally accepted y is accepted until it is shown to be false. But if the null hypothesis is false, the probability that the hypothesis is true is negligible. The scientific method is the best known way to obtain knowledge with a high probability of being true. In agronomy, it is particularly important to experiment in the field, since it is practically impossible to simulate, to date, in laboratory experiments, growth chambers, phytotrons or greenhouses, the conditions that occur in the field. The profiles of wind, light, temperatures, CO_2 of a crop during the day and at night in these enclosures are very different from those in the field. The field experiment is, so to speak, the final test of many agronomic hypotheses such como lo are the techniques, which we can consider como hypotheses, proposed to solve some problem.

Thus, hypotheses relating to the effect on plant populations of: mineral nutrition, the effects of water, genomes, disease control, pests, soil tillage, conservation of the harvested crop, etc., can be tested using this method. It also provides a quantitative measure of the effect of the factor under study. Only by means of field trials is it possible to give an agronomic verdict on a hypothesis. The results of off-field experiments are only provisional and have the advantage of being more controlled, cheaper o faster o and if they are positive they are a good indication that the hypothesis tested may be true in the field but they do not guarantee it. If the results are negative, they indicate that the hypothesis is not correct and a field test is unnecessary. A limitation of field testing stems from the time it takes to conduct a trial, usually several months, which reduces the number of hypotheses a researcher can test.

On the other hand, to gain a deeper understanding of what is happening at the physiological or soil or other levels, it is necessary to carry out laboratory studies which are more appropriate to these levels of complexity. This means that any organisation seriously studying agronomic problems must have adequate facilities for research at both the field and laboratory levels.

In general, the field trial consists of sowing plots, with a minimum of three replicates established at random, which are treated with one or some of the factors to be studied and others that receive no treatment, which are called controls. The treatments generate some variability in the properties of the plants in each plot, e.g. yield. The other uncontrolled factors are assumed to affect randomly and have the effect of increasing the variability of the measurement made y affect all plots equally. When it is known how an uncontrolled factor varies spatially the replicates are distributed in blocks. Essentially it must be established whether populations receiving different treatments are different o not from controls. Analysis of variance allows to calculate the variabilities caused by the controlled factors, by the blocks o by the uncontrolled factors o by the uncontrolled factors o by the treatment effect, block effect o and "error" effect. One way to know if the treatment effect is significant is to calculate the ratio between the variance due to the treatment and the variance due to error. If their ratio is greater than a certain given value, provided by F tables, according to the probability y to the degrees of freedom, it follows that the populations that received treatment are different for that probability and the treatment effect is significant. This test is valid if the variables, measured in the plots, are normally distributed. To find out if this is the case, a normality test o a histogram of the distribution of the data is made. If the histogram is close to a сатрапа-type curve, it is normal. The exact form of a normal distribution is defined by two parameters: its mean and its standard deviation.

Probably the main problems that make it difficult to interpret the results of these trials are the influence of and interactions with the dynamics of meteorological and biological factors, spatial variability of soils and sanitary conditions, and climatic factors that vary from year to year. The annual effect makes it desirable to repeat the same experiment for at least three years before any defensible conclusions can be drawn.

Ideally, it is feasible to postulate that if one could simultaneously measure, over an area, the main factors that determine the behaviour of each plant in a population, it would be possible to find a function that describes the variations in yield as a function of these variables. The use of measurements of soil factors and plant biophysical properties via multispectral image analysis provides current researchers with a way of making many measurements of variables in space and time, with a spatial resolution that is practically impossible to make in other ways. Elio, eventually, could be a new alternative for advancing agronomic knowledge.

Unfortunately, these days, agronomic field research has been regarded as сото of less scientific value y is becoming more and more difficult to finance. This is a tremendous mistake which I hope will be corrected in the future. Without field research there is no adequate support for practical agronomy.

We know that modern agronomic knowledge has come from the use of the scientific method. This method is essentially a useful сото analytical tool but a theory requires taking this information and reconstructing the whole system under study. This is perhaps because the human brain works in this way, the information it receives from the senses is first analysed and then reconstructed, Kandel, 2006.

This integrative approach, as opposed to the analytical approach, is first found in multivariate static models to estimate quantitatively the effect of various physiological and environmental factors on photosynthesis, biomass production and yields of some crops (Saeki, I960, Loomis y Williams, 1963, de Wit, 1965, Duncan et al, 1967, Chartier,1966,1970) and, subsequently, in dynamic crop simulation models (de Wit, 1970a,b, Fick, Loomis y Williams, 1975, van Keulen, 1975, Baker, et al. 1979, Novoa and Loomis, 1981b, Novoa 1987, Richtie et al.,1985, Thornley and Johnson,1990, Liu et al, 2007a,2007b, Yang et al. 2009, Steduto et al. 2009 y many others). Statistical models were also developed that also sought to identify and quantify the effect of factors affecting crop production. This does not mean that the integrative approach displaces o replace the analytical approach, called reductionist by some, but rather the idea is that agronomic science, essentially integrative, advances by taking advantage of the results of both approaches.

Agro-ecosystems are complex systems and therefore possess the typical characteristics of such systems: the whole is more than the sum of its parts; they have emergent properties; they evolve; they are open dissipative systems in which energy and matter flow; they function outside equilibrium, which requires continuous sources of energy; their interrelationships are governed by non-linear equations; their behaviour is difficult to predict and therefore statistics are used to make predictions. On the other hand, natural ecosystems are self-organising. If we consider man сото an integral part of the system, agroecosystems are also self-organised. A basic difference between agroecosystems and ordinary ecosystems is that in the former it is possible to identify the organising entity, while in the latter it is not. In agroecosystems, man gives them stability, the ability to learn in an organised way and to evolve faster.

In the opinion of Fisher, 2007, one must not lose sight of the field environment, the bottom-up approach (from the molecular upwards) is challenged by complexity and greater emphasis is required on top-down approaches that lead to the establishment of key mechanisms. The Hamada emergent properties support this view. These properties appear when various elements are combined in a system y are not deducible to the properties of the constituent elements, but probably result from their mutual interactions. A simple example is that of water, by combining two gases, O_2 y H_2 , a molecule with the properties of a liquid emerges. Another simple example is that of the response, positive or negative, that emerges when a flow of information operates in a feedback loop (an increase in photosynthesis normally produces leaf growth and this results in an increase in photosynthesis). Thus, the behaviour of a plant cell is different from that of a plant and the behaviour of a plant population is different from that of a plant in the same environment. Emergent properties create increasing constraints o new conditions as the level of organisation increases. This makes it difficult o incomplete the bottom-up approach o and at the same time explains como the randomness o degrees of freedom, which govern the subatomic o atomic world o atomic world is reduced making possible the emergence of laws governing the behaviour of higher levels. However, this does not imply that all the properties of a population are emergent, since it is possible to explain to a large extent many properties or behaviours of populations como results of molecular processes (photosynthesis, respiration o transpiration como processes of gas diffusion, CO_2 y water, for example) plus the effect of biological limitation (resistances: boiochemical, mesophyll, stomatal, leaf boundary layer y crop, diffusion). Similarly, the physical and chemical laws governing energy, atoms and molecules cannot be violated. On the other hand It seems clear that the behaviour of complex systems, whose nodes are usually connected to each other, is determined by important components that drive the system. They are variable, key nodes of the network that determine the state of the other nodes. Recent work by network researchers has developed a technique for identifying these variables and tracking the behaviour of complex systems by monitoring only these variables over time (Liu et al., 2013). It is possible that this technique will allow the identification of key agronomic variables in agricultural ecosystems.

From this perspective, agronomy would be the science in charge of identifying, studying the emergent properties and establishing the laws of agro-ecosystems, while genetics, physiology, biochemistry, chemistry and physico-chemistry are best suited for molecular studies. To a large extent, the emerging properties of agro-ecosystems, established to date, are summarised in the principles mentioned below.

Agricultural systems, integrating physical, chemical and biological, human and socio-economic factors, are of a higher level of complexity than the purely chemical, physical and biological. They must function according to the laws and constraints that apply in each of these areas, plus the information flows (e.g. feedbacks) and dynamics that are specific to their level of complexity. This should not be forgotten when studying these systems. Thus, a chemist, without agronomic or ecological training, who studies the use of fertilisers must be aware that his chemical knowledge is not enough, since the influence of soil myco-organisms and plant roots, of man via soil tillage, fertiliser application times, waste management, irrigation, etc., substantially modify what happens to the ions in the soil. On the other hand in the field both soil como sanitary conditions, hydric (rainfall y soil hydric properties), atmospheric (CO2 concentrations, temperatures, solar radiation, humidity) are variable in space (horizontally y vertically) y time. This generates gradients that do not occur equally in laboratories but must be managed in practical agronomy. These tests give clues about possible mechanisms involved in the observed responses but are not a tool that can replace experiments aimed at establishing the more basic physiological, biochemical or other processes.

Numerous studies have demonstrated the high economic return on investment in agricultural research (Evenson et al, 2003a). One difficulty in estimating the cost-effectiveness of a particular research project is that it is very difficult to estimate its real impact when doing ex-ante analysis. It is practically impossible to know a priori which researcher is going to make a high impact contribution, but what is known is that if one considers the whole body of research, there will always be one result that is profitable and will pay for all the failed attempts. The system works in the reverse of the insurance business where the many pay the claims of the few. In research the few successful ones pay for the investments in the many.

Modern agriculture is based on the application of the principles and laws described below.

Chapter 2

2. - PRINCIPLES[1] AGRONOMY

Interestingly, when reviewing various agronomy texts, it is possible to note that they mention agronomic principles, but the usual thing is that they do not specify what these principles are. However, modern agronomy is based on the application of the following three principles (Novoa 1986, 2004):

2.1. - Use of organisms, selected o improved, efficient. Agriculture makes use of populations of living organisms to produce the desired goods. In a sense these represent the machines used to transform solar energy and inorganic chemical elements of their environment into goods of human interest. To a large extent the productive efficiency and the quality of the product are determined by the organism used to produce. Thus, varieties or breeds (genetic factor) and/or modifications of the organism used to improve its efficiency (surgical techniques, pruning, grafting, use of hormones and others) are the options available to achieve high yields and optimum quality. One of the main tools used to improve the productive efficiency of the organisms used in agriculture has been genetics. It normally takes between 7 and 20 years of research to make an improved organism available to farmers, using traditional techniques, which implies keeping these programmes active for many years before they start to have an impact on production. Improvements in annual crop productivity due to breeding have been estimated at 0.83% during the 1980s to 1990s but the impact has not been equal in all countries and regions. In the case of wheat these improvements have been brought about by improvements in resistance to dusts and in the harvest index brought about by the introduction of dwarfing genes (Austin et al, 1980). At the same time consumers have benefited from lower prices and farmers have only benefited when cost reductions have been greater than price reductions (Evenson, 2003a, b).

Genetic and epigenetic factors confer an intrinsic efficiency on the organism used to produce.

The genome, DNA, of an organism defines the genes that characterise that organism. They establish the sequence of amino acids that a protein can have. Proteins can be considered nanomachines that form a complex network which gives rise to the functions of the organism, they are in charge of doing the work in the cells. Thus, we find them fulfilling the functions of catalysts in the case of enzymes, structural, mass transport (ion channels o O transporters[2] , сото haemoglobin) o information (hormones). In addition, they can act on their own o forming complexes in which two o more proteins are joined together to perform a function.

However, a gene can give rise to more than one protein, and one of the processes by which this phenomenon occurs is the modification of genetic information called splicing. In the case of RNA it may be a post-transcriptional modification and in the case of DNA the covalent joining of two DNA fragments. The information to make a protein is organised into blocks of DNA called exons separated by other DNA sequences called introns whose function is not known. The modification process, which removes introns, can follow different pathways and thus generate different proteins. For example, removing one, or two, amino acids from the sequence. In addition, the diversity of proteins may arise from post-synthesis changes to the amino acid sequence such as o addition of phosphate groups o sulphates o addition of carbohydrates.

[1] The Royal Spanish Academy defines principle сото "each of the first propositions or fundamental truths from which one begins to study the sciences or the arts" and a law сото "each of the existing relations between the various elements that intervene in a phenomenon" _.

On the other hand, in the human genome, it has been shown that genes are generally found in separate sections located in different places on the DNA molecule, which requires assembly for translation. The same gene can be assembled in different ways, a process called "alternative splicing", giving rise to messenger RNA molecules and proteins that have different functions. This mechanism enhances the genome's capacity and provides it with great flexibility, Ledford, 2008. It is also known that only 1.5% of human DNA encodes proteins while the other 98.5% contains regulatory sequences that activate or silence genes, genes that encode RNA but do not produce proteins and other DNA whose purpose is still being understood (Pollard, 2009).

In addition, the expression of a gene can also be modified by the action of one or more genes, a phenomenon known as epistasis.

The fundamental importance of genetic factors is increasingly well understood thanks to advances in the science of molecular genetics. Molecular genetics has provided geneticists involved in the breeding of species of interest with powerful tools that allow genes from any species to be inserted into the genome of the species to be bred. This makes it possible, in theory and with greater certainty than with traditional techniques, to obtain improved varieties or breeds with high yield potential, good quality, good resistance to diseases, pests or environmental stresses.

The general view is that genetics defines the productive potential which is modulated by ecological and management factors. In other words, the maximum yield is defined by genetics, if there are no ecological or management constraints. Genetics defines the species: its photosynthesis system (C3, C4 o CAM); the organ of agronomic interest, which is usually the organ where the plant stores its reserves; its fruit (e.g. if it is the fruit, it can be a berry, it can be a berry): if it is the fruit it can be berry, drupe, nut, achene, caryopsis, legume, or other), , the shape of the plant, the type of leaf (broad o thin, with o without trichomes), its photoperiodic requirements, its cardinal temperatures (minimum o base, optimum y maximum), the base temperature for growth, the thermal requirements of each stage of the life cycle, the shape and location of stomata, resistance to disease and pest attack, the type of roots (pivotal, fibrous, adventitious), roots with the ability to form symbiotic associations with rhizobium, roots with aerenchyma that can withstand waterlogged soils, tolerance to salinity. While the size, colour, chemical composition of organs at each phenological stage of growth, harvest index, leaf area, leaf angle, life cycle and length, and other yield-determining characteristics are under genetic and environmental control. The possibility of increasing yields by this means is still clear (Mangelsdorf, 1966) since in general there is genetic variability for many of the genes involved in yield and with the advance of molecular genetics the number of genes that can be exploited for this purpose is large.

In the case of epigenetic inheritance, information is transmitted without modification of DNA. Epigenetic factors are non-genetic mechanisms that reversibly modify a gene, or the proteins it gives rise to, by activation or inactivation of genes as a result of interaction with the environment. This modification can be by gene silencing o via DNA methylation, o phosphorylation o histone acetylation for example. They provide a mechanism by which the genome responds to environmental changes. Thus some species of watercress do not flower in winter because of a low temperature action on chromatin that silences the flowering genes. These genes are reactivated by higher spring temperatures. Epigenetic processes modulate the action of DNA without modifying it. Epigenetic information can be visualised as an engraved code outside the gene that turns a gene on or off. In cereals the response to vernalization is mediated by induction of the Vernalization 1 promoter that initiates reproductive development of the stem apical y is associated with histone methylations in the Vernalization gene (Oliver et al, 2009).

As genetics establishes the productive potential of an organism it is very important to know what this is. One way to get an idea of it is to know the yield potential of a crop. This is defined coтo the yield of a cultivar when grown in a parallel environment which is adapted with water and non-limiting nutrients, with pests, diseases, weeds, stresses and other stresses effectively controlled (Evans,

1993, Evans and Fischer 1999). Another, more functional, definition is the yield obtained when an adapted cultivar is grown with the lowest possible stresses that can be achieved with best management practices, Cassman, 1999. This is a fundamental data whose importance has been recognised by Mitsherlich, 1909, De Wit, 1965, Doorembos y Kassam,1979, Hanson et al, 1982. Not only because it determines the maximum possible yield, but also because it allows us to visualise the economic income potential and thus to infer up to what point it is economical to invest in technologies to increase yields. Thus, if the maximum yield is 12 tonnes of grain, it allows for greater expenditure on technology than if it is only 6 or 8 tonnes. There are scattered data on the maximum yields obtained for different crops in different countries by farmers and researchers. Table 1 shows the maximum yields measured y the world averages produced and it can be seen that there is great scope for improving the averages. It should be borne in mind that the maximum values obtained are valid for the genotypes, ecological conditions and management in which they were measured and cannot be believed to be replicable elsewhere. Nevertheless, they give an idea of the great genetic potential of each genotype studied.

Attempts have been made to experimentally measure the value of yield potential in the field but given the effect of climatic, soil, sanitary and other factors, it is practically impossible to achieve ideal zero restriction conditions that allow the full genetic potential of a variety to be expressed. Thus in the case of wheat there are reports of field yields of 15.6 ton/ha in New Zealand (cultivar Einstei, FAR 2010), 18 in Chile Hewstone, 1997, y 47.6 in growth chamber with the spring cultivar Yecora Rojo (Salisbury, 1991). The latter yield was achieved with 2000 plants m^{-2} , CO concentrations$_2$ 3.6 times normal, radiation 2.5 times higher than what the field plants could have received, no water limitations, optimal nutrition, no diseases or pests y with an optimal thermal regime. The harvest index was 0.49. This author found that the efficiency of photosynthetically active radiation use decreases as the intensity of radiation increases, that in the period of maximum growth it was almost 13%, and that for the complete cycle, the efficiency obtained was only 10%, a measurement that corroborates theoretical estimates of this efficiency (Loomis and Williams, 1963, Taiz and Zeiger, 1998). The maximum yields obtained are much higher than the record yields obtained in the field in Table 1. Salisbury attributes the lower yields measured in the field trials to possible stresses, lower radiation and lower plant density per unit area.

The use of stress-free growth chambers seems to be able to approximate the genetically possible yield potential como Salisbury will show. Another promising alternative for establishing yield potential is the use of dynamic simulation models, but to date, these do not allow calculation for every existing variety and are not easy to use. This is due to the complexity of the models, the numerous parameters required, which are not known for all varieties, and other limitations.

Table 1. Measured maximum commercial product yields y world averages*, ton ha-[1] .
Table 1.- Maximun measured and world average yield of agricultural goods, ton ha'.[1]

Cultivation	Global average performance'.	Variety o breed	Maximum Performance	% water	Region y Country	Referenda Maximum yield
Rice	4.2	II-32A/Ming86	17,95	11	CHINA	Longping, 2004 Bayer Cropscience, 31 August 2005
Wheat	3.0 2	Savannah	15.36		New Zealand	http://www.fwi co.uk/community/blogs/arableb arometer/archive/2007/11/01 /world-record-wheat-yield-exposes-growing-constraints.aspx 2007
Wheat	3.0 2	Yecora Red	47, 8		Utah, USA	Salisbury, 1991
Maize	5.1		23,2	13	USA	Duvick and Cassman, 1999
Sorgo	1.4 2					
Barley	2.7 7		11.4	11		Evans, 1993
Soy	2.2 5	Pioneer 94M80	10.36	9	USA	http://www.pioneer.com/web/site/portal/men ui tem.b5db62e333774bbc81127b05d10093a0 / 2007
Sugar cane	70.9 1	Co62175	11 0	70	Chittoor, INDIA	http://www.atmachittoor.com/SS-Sugarcane.htm
Sugar beet	53.0 8		157.2	75	California, USA	Rush et al. 2006
Potatoes	17.9 8		44.39	78	Nebraska, USA	Paviista, 2002
Casava	12.6 4		40.9			**Nair and Unnikrishnan**
Mani	.49	ICGV 87123	5.76		East Timor	Nigam et al. 2003
Tomatoes	27.2 9		580 3	94	MEXICO	Cook, 2006.
Banana	19. 7		60	75.5		http://www.research4development.info/Sear c hResearchDatabase.asp?ProjectID=2464
Coconut	5.4		7		Malaysia	http://diversiflora.blogspot.com/2006/01/des cr iption-matag-hybrid.html
Alfalfa			54'		USA	USDA, 2000
Oats		S-81	69.4	10	Faisalabad, Pakistan	Naeem et al 2006
Milk		Jersey	22330		USA	http://jerseyjournal.usjersey.com/Article %20Archive/Production/2007/Jubilee %20 Article.pdf
Eggs			361[5]		Netherlands	http://www.timecom/time/magazme/article/0, 9171,848100.00.html
Came bovine	205 7		4300[8]			

1,- Yield in the field 2,- Growth chamber 3,- Greenhouse *4,-* Kg per year per cow 5,- Eggs per hen 6.con irrigation 7,- kg meat per animal 8,- kg ha'[1] estimated max. alfalfa production per ha, loss 20 % trampling y requirement 10 kg alfalfa per kg meat
* According to FAO 2007-2009

If the yields are corrected for moisture content, C3 plants produce a maximum of between 9 (potatoes) and 39 (sugar beet) tonnes of dry matter per ha, C4 plants between 20 (maize) and 33 (sugar beet) and leguminous plants between 10 (soya) and 25 (alfalfa). On the other hand, to determine the total biomass production capacity of each species, it is necessary to know the harvest indices y if one wishes to express the production capacity in terms of photosynthesised glucose per unit area per cropping season, one must correct for the chemical composition of this biomass (Sinclair y de Wit, 1975). The high maximum yields measured suggest that the available genetic potential is much higher than is generally thought and that better agronomy is required to exploit it properly.

Tom Sinclair, in 1994, stated that all the information known up to that time indicated that the limit of yields possible under good management had been reached. It stands to reason that at some point this will inevitably happen. Cassman, on rice 1999; Rajaram y Braun, 2006, on wheat in the Yaqui Valley, Mexico; Carberry et al. 2010, on rainfed maize in Australia, provide evidence of this phenomenon. However, FAO data on world averages of wheat, rice, soybeans, barley, maize and sorghum between 1960 and 2007 show yields rising with a slope that shows no evidence of saturation except in the case of sorghum. There is also no evidence of this according to studies done on wheat yields in North Dakota, USA, (Underdhal et al., 2008). Other more recent analyses show that the situation depends on the country. In general there is a decrease in the annual yield increase of wheat, maize, rice and soybeans. In industrialised countries the annual yield increases in wheat in recent years have reached zero while in the case of maize the decrease in yield increases has been smaller. On the other hand, if the linear growth in annual yields observed over the last 50 years of 43 kg ha^{-1} is maintained, the future demand for cereals for food, feed and biofuels cannot be met (Fisher et al. 2009). On the other hand, there is clear information that agricultural production continues to grow and that total factor productivity (TFP) has increased (Fuglie, 2008, 2009).

Although, the relationship between photosynthetic rates and crop productivity is not clear (Evans, 1975) as it would be more limited by storage capacity (grain) than by any other cause (Borras, Slafer and Otegui, 2004), it seems possible to improve photosynthetic efficiency. Analyses of the radiation use efficiency of wheat and maize indicate that there is scope for increasing the yield potential of many crops, Loomis and Amthor, 1999. Given that the current efficiency of the photosynthesising system in intercepting solar radiation is close to the theoretical limit, the harvest rate of cereals in the order of 60 %, close to the maximum possible limit (Hay 1995, Austin et al. 1980), it has been suggested that improving the efficiency of converting trapped light energy into biomass would be the most suitable route to improving yields in the future. It has been estimated that the yield potential could be increased by 50 % como as a result of the combined effect of the improvements summarised in Table 2 (Long et al, 2006).

Table 2. Summary of possible increases in solar radiation conversion efficiency (εc) that may be achieved and the speculated time horizon for provision of material that can be introduced into plant breeding programs. Summary of possible increases in solar radiation conversion efficiency (εc) that may be achieved and the speculated time horizon for provision of material that can be introduced into plant breeding programs.

Change	% increase in £c relation to current values[ll]	Ho rizonsfrom estimated times (years)
Rubisco with lower oxygenase activity y without lower catalytic activity	30% (5-60%)	???
Photosynthesis C4 introduced in C3 crops	18% (2-35%)	10-20
Best canopy architecture	10%(0-40%)	0-10
Mayortasade recuperation by photo protection dela photosynthesis	15% (6-40%)	5-10
Introduction of Rubisco forms[2] with increased rates of CO_2	22% (17-30%)	5-15
Increased regenerative capacity of RuBP via overexpression of the SbPase[3]	10% (0-20%)	0-5

[2]Rubisco: ribulose 15-biphosphate carboxylase/oxygenase; RuBP, ribulose biphosphate.

[ll]The value under the heading '% increase' is the suggested mean, followed by the range of possible changes calculated by substituting the suggested changed properties in the simulation model of Humphries and Long (1995).

[3]for Sbpase: sedoheptulose-1:7-bisphosphatase

The biomass formation potential could be doubled if the breeding suggested by Long et al, 2006, were to materialise. On the other hand, yield studies of super rice hybrids indicate that their advantage would be in the growth rate in the middle and late stages after flowering due to the effect of higher leaf area index and longer duration of leaf area Wen-ge et al, 2008.

There are projections of possible increases of 50% for most crops by 2050 (Jaggard et al. 2010).

A projection of corn yields towards 2030 gives a possible yield of 18.8 t ha^{-1} (Calabotta 2009). Monsanto projects corn yield increases to double by 2030 due to 25% due to improved agronomy, 35% due to traditional breeding, 25% due to transgenesis and 15% due to selection using markers (Edmeades et al, 2010).

The case of the enormous genetic potential of wheat shown by Salisbury shows that, at present, we are not limited by genetic potential and that if we improve agronomy and management, the food capacity of agriculture can be increased very significantly. Thus there is strong evidence that the productive potential of a crop can be raised. Furthermore, it seems possible to approach the yield potential of a population of plants if we can provide optimal conditions for each plant in the population. One possible approach to partially solve this problem, in the field, is site-specific or precision agriculture, which seeks to manage the spatial variability of soil, health, nutritional and water conditions.

The effect of breeding, in the case of wheat, has not been due to changes in the productive potential of photosynthesis but to effects on the harvest index by introduction of the Rht1 and Rht2 genes that reduce the height of plants provided by the Japanese variety Norin 10, disease resistance сото dust or other y resistance to lodging (Austin et al, 1980, Hanson, Borlaug y Anderson, 1982) y in the case of maize to increased tolerance to high planting densities, due to more erect leaves that reduce shading between plants y increased ability to retain their green colour (Evans y Fisher, 1999, Loomis y Amthor, 1999).

The harvest index (0.4to0.6 in the case of cereal grains) considered to have high heritability and sensitivity to environmental factors (Hay 1995) is however affected by ecological factors such as N availability (van Ginkel et al, 2001) or temperatures. Thus, Chilean farmers speak of the "ida en vicio" that N o high temperatures can produce, referring to the increased production of leaves and stems to the detriment of grain production in cereals and fruit trees. However, this effect does not always occur, which implies the effect of another unknown factor. The harvest index can also be affected by terminal water deficit or pest and disease attacks during the period of formation of the crop to be harvested.

2.2.- Provide suitable ecological conditions. The genetic potential of the organisms chosen can only be expressed if the ecological conditions that allow it to be managed are adequate. The climatic, atmospheric, soil, water, sanitary (pests, diseases, weeds) and nutritional conditions that make up the ecological environment (oikos, the word from which ecology derives, means house in Greek) affect the productive processes and the development of the population of organisms used. A large part of agronomy consists in providing the best possible ecological conditions for the plant or animal species used.

2.2.1. - Climate and weather conditions.

Climatic conditions represent the evolution of the average of the micro-meteorological conditions (defined as those physical characteristics o occurring in the first 10 metres of the atmosphere) that are to be expected at a given location over a period of time.

2.2.1.1. - Electromagnetic energy conditions. The microclimate provides the regimes of short-wave electromagnetic energy: ultra violet below 400 nm, visible o photosynthetically active (PAR) between 400-700 nm and infrared, above 700 nm, which will be received by the cultivated plant population. PAR is absorbed by a complex of pigments including carotenoids, chlorophyll a y b that create an "antenna" in photosynthesising tissues, mainly leaves, to generate a reducing power that is used to decompose water, reduce CO_2 y to form glucose. From glucose, amino acids, lipids and other components of plant biomass are formed. The sum of the PAR to which a crop is exposed during its life cycle sets a maximum limit to the production of plant biomass, which is reduced by plant, atmospheric, hydraulic, sanitary and other factors.

The blue and red radiation captured by photoreceptors, consisting of an apoprotein and a chromophore, produces phenomena called photomorphogenic. On the other hand, blue radiation (400-500 nm) perceived by photoreceptors produces phototropism: phototropism, stimulating bending towards the light by producing an uneven distribution of auxin in growing organs, which is greater on the shaded side; inhibition of stem growth; regulation of the expression of genes controlling the synthesis of chlorophyll and flavonoids; stimulates stomatal aperture via regulation of guard cell osmotic relations by activating a proton pump in these cells (this response follows the law of reciprocity, i.e. it responds to light dose, not flux or exposure time) with the carotenoid Zeaxanthin being the photoreceptor involved. The red (666 nm) and far-red (730 nm) radiation is detected by the phytochrome changing its form from Pr to Pfr. The Pfr form that absorbs far-red light is the active form that produces the following responses: the transformation of oat seedlings from etiolated to green under the effect of light; controls the photoperiodic response of various plants; promotes germination in lettuce; controls the transcription of the LHCB family of genes encoding the chlorophyll-binding proteins of photosystem II involved in the development of green colour; represses the transcription of the LHCB family of genes encoding the chlorophyll-binding proteins of photosystem II involved in the development of green colour; it also represses transcription of several genes including PHYA encoding the phytochrome A apoprotein; it alters membrane properties and flowering, Taiz and Zigers, 1998. In the case of flowering, a distinction is made between long-day (barley, oats, wheat, clover, lilium spp., Beta vulgaris, spinach, cabbage, radishes), short-day (rice, some strawberries, some tomatoes) or neutral-day (maize, Phaseolum spp., potatoes, tomatoes) plants, depending on whether the dark period (which accumulates Pfr) has a critical duration or not. Some animals сото sheep, beef cattle (Yeates, 1955,1957, Glover, 2000) o dairy (increased milk production on long days) o hens (increased egg production on long days), are also affected by photoperiod (Dahl et al, 2000).

On the other hand, the nocturnal infrared radiation balance between the atmosphere and the atmosphere determines the air temperature. The band 8000 to 11000 nm plays a determining role in this balance, it produces the so-called "atmospheric window" because the atmosphere is transparent to radiation from this band. The energy released by the radiation of this band is lost by the earth. As the energy input from the atmosphere to the ground is poca, the radiative balance at night is normally negative, the air and the ground cool down. When this balance is very negative the cooling is great and temperatures below 0 °C o freezing. This condition can adversely affect crops, especially at very sensitive stages such as flowering.

2.2.1.2. - Thermal conditions.

The thermal conditions of the atmosphere and soil are determinants of the growth and development of plants and animals.

In general, metabolic reactions increase in rate with increasing temperature because the temperature increases the kinetic energy of the molecules, which leads to an increase in the collisions between them. In biochemistry, the so-called Q_{10} has been defined, a factor that gives the increase in the rate of reactions for an increase of 10 °C, this increase is 2 to 3 times depending on the reaction considered (Nobel, 1974). It is usually of the order of 2.2 for the case of vegetates y temperatures between 10 y 30° .

Temperature has a great influence on germination, growth and metabolism of plants. If it is too low there is no germination and no growth. Above this minimum there is a positive effect on growth and metabolism and above a maximum there are negative effects.

Temperature affects the growth of all living organisms. Temperatures below - 0.5 °C produce damage typical of frosty days to sensitive tissues сото flowers, due to the formation of ice crystals that break the cell walls as they grow. Normally, each organism functions (grows and develops) according to a base temperature (Tb), an optimum (To) and a maximum (Tm). In the case of cultures

Most plants below 0° C there is practically no germination or growth. However, depending on the type of plant, growth starts when the temperature rises above the critical value for each species. The lower values, around 5 °C, are plants that grow preferably in winter and the others (8 to 15 °C) in summer and above 18 ° C tropical plants. The maximum tolerated temperature is between 35 y 45 °C depending on the species. Temperatures also affect the rate of development following a relationship shown in Figure 1 for maize (Craufurd and Wheeler, 2009).

Figure 1,- Relation between the rate of development y the temperature
Figure 1,- Relationship between development rate and temperature

Mean temperature (°C)

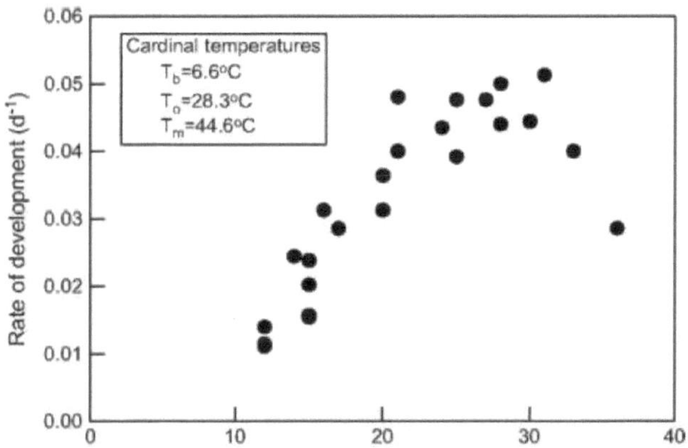

The life cycle of a plant is shortened by increasing temperatures because increasing temperatures increase the rate of plant growth. This reduces the length of the growing season and the radiation absorbed, with a consequent reduction in total net photosynthesis. Thus, in climates with warm nights, respiration during the night is much higher than in climates with cool nights and net photosynthesis is lower. This explains why in Chile and other countries with warm days and cool nights higher yields are achieved than in countries with warm days and nights.

The concept of daily temperature sums (TS) that a plant needs to accumulate to complete a phenological period has been widely used and discussed. This sum is calculated according to formula 1 by subtracting from the average temperature measured in a meteorological shed the base temperature specific to each phenological stage of development. At 17

In principle, this sum would be specific to each state of each species and plant variety.

ST = Z(Tmeans-Tb) (1)

If this sum is known for each stage, it is possible to define the days during which 50 % of the plant population, in a given climate, will take to complete the corresponding stage. Thus, sowing-emergence, emergence- tillering, tillering-flowering, flowering-maturity or other days can be estimated. Obviously, the calculated values are approximate, as there are other factors that can affect the development of the crop, such as photoperiod, nutritional status, water status and sanitation. Their use is criticised for being based on the average temperature value which can be the same in very different conditions. A maximum temperature of 3O and a minimum of 0 gives a mean temperature of 15 as does a maximum of 18 and a minimum of 12, the two regimes being very different.

In addition, there are plants which are stimulated to flower by temperatures below 7°C, a process called vernalization. In the case of winter wheat, it requires about 800 hours below 7 for good flowering.

In the case of fruit species such as plums, some varieties of peaches and sour cherries, cherries, olives, apples and pears require exposure to cold periods in order to flower and therefore their production decreases if they do not have it.

Table 3 summarises the cooling requirements of some fruit trees.

Table 3: Cooling hours requirement of deciduous fruit trees (n° hours <7°C)

Table 2: Cold hours requirements ofdeciduous fruits trees (hours< 7°c)

Species	Minimo	Maximp
Almond tree	100	500
Blueberry	700	1200
Hazel	800	1600
European plum	700	1600
Japanese Plum	100-600	1000
Damascus	200-500	900
Peach tree	100-400	1100
Guindo	600	1400
Cherry	500-800	1500
Kiwi	800	1400
Manzano	200-800	1700
Membrillero	100	500
Nogael	400	1500
Pecano	600	1500
Peral	500	1500
Vine	100-500	1400

Source: Flores, 2007

2.2.2 Atmospheric conditions. The atmosphere provides mainly CO_2 , water vapour, N, thermal and aeolian environment.

All the carbon that forms plant biomass comes from atmospheric CO_2 whose concentration in the air influences photosynthesis and has a "greenhouse" effect by reducing energy losses via longwave radiation (around 10,000 nm) from the earth. Human activity has generated an increase of this gas in the atmosphere, which leads to an increase in photosynthesis by increasing the gradient of this gas between the sub-stomatal cavity and the atmosphere.

Precipitation is the main source of water to agricultural soils, while the water vapour content of the atmosphere affects evapotranspiration, the higher the water vapour content, the lower the evapotranspiration. Higher humidity also favours the attack of fungi. Thus, the positive effect of dew, atmospheric water condensed on the leaves, on yields is somewhat reduced by a higher incidence of fungi (Novoa and Letelier, 1987).

In the case of most soils a large part of the N comes from the symbiotic and non-symbiotic fixation of atmospheric N. Another part of the applied N como fertiliser and another part from the decomposition of organic matter.

As for wind conditions, wind is a factor that influences the diffusion and mixing of atmospheric gases, the temperature of both the air and the surfaces over which it flows, transfers momentum to the vegetation and the ground (Thom, 1975). It affects the transport of CO_2 y water vapour y hence evapotranspiration.

Air movement replaces the CO2 extracted by plants, transports evapotranspired water vapour and attenuates thermal extremes. On the other hand, very high velocities can lead to the stretching of crops como cereals o breakage of branches on fruit trees. At lower speeds, between 0 y 7 km hr^{-1} , the wind speed affects the bulk of the boundary layer (sector where the flow is laminar, without turbulence, where the movement of gases is only by simple diffusion due to molecular motion) of the leaves. Above that velocity turbulence is produced and the flow of gases is increased because they move in sort of "bubbles or vortices" that are formed and the gases travel in them much faster than by simple diffusion. In turn, the movement of the leaves facilitates the formation of turbulent transport. It also transports clouds, sand, fungal spores and seeds.

2.2.3 Soil conditions.

The properties of a soil: its slope, stoniness, depth (impermeable layers, water table, bedrock), granulometry or texture, porosity, pH, salinity, drainage, fertility, organic matter content, presence of weed seeds, fungal inoculum, nematodes and others affect the growth and development of plants.

The soil basically provides:

2.2.3.1. Sustainability for plants and animals.

2.2.3.2 Reserves of: thermal energy, water, nutrients (N, P, K^+ , Ca^{++} , S, Fe^{++} , Mg^{++} , Mn^{++} , B, Cu^{++} , Zn^{++} , Mo, Cl^- ,Ni^{++} y the necessary Si in cereals como wheat o oats y the Na^+ in beet y plants C4) y germplasms both vegetates (seeds) como animates (eggs). Normally, the main nutrients (N, P and K) in the soil are deficient, with less frequent deficiencies of secondary nutrients (Ca, S and Mg) and rarer deficiencies of the other elements (B, Mn, Zn, Cu, Mo, Cl, Ni and Si), which are called minor. This is why N fertilisation is always necessary, followed by P and then K fertilisation.

2.2.3.3 A habitat for a wide variety of organisms: bacteria, fungi, algae, plant roots and a variety of micro- and meso-fauna. These organisms make the soil more than just a collection of inorganic materials subject only to chemical and physical laws, since to these are added biochemical and biological properties that modify their behaviour. This reality is not always understood, but it is very evident in the case of nitrogen, the availability of which to plants varies throughout the year as a consequence of the biological activity of the soil. A process of enormous impact on ecosystems and agriculture is that of atmospheric nitrogen fixation. Particularly if we consider that practically all the N to which terrestrial plants had access, until the 20th century and to a large extent until today, came from this fixation (Zagal, 2005, Roy et al, 2006, E. Letelier personal communication). In fact the minerals that form the parent rocks of the soil do not contain N. The process of atmospheric N fixation that is carried out by bacteria (such as the legume synbionts *Rhizobium or* the free-living *Azotobacter* Cyanobacteria) and blue-green algae is comparatively similar or less important than photosynthesis. Of the 275 million tonnes fixed annually, it is estimated that 175 million tonnes are biological fixation, 70 million tonnes are industrial processes and 30 million tonnes are spontaneous fixation (Rodriguez et al. 1984). Among the N-fixing bacteria those that perform the bulk of this transformation are the *Rhizobium,* bacteria that live in the roots of plants of the Leguminosae family. These bacteria require energy, in order to reduce atmospheric N_2 molecules, which is provided by carbohydrates that the plants generate in the process of photosynthesis. By providing N to leguminous plants, *Rhizobium* is increasing the availability of an element, present in very limited quantities in soils, which is essential for both the formation of protein coтo for living organisms. Although it is true that N is fixed in the roots of leguminous plants, it passes into the soil when the roots and other organs of these plants die. In this way, the N fixed in this way is available to plants of other families, improving their growth and development. The amount of N that can be fixed by an alfalfa crop can be deduced from the crop's potential yields, its N content and the soil input. Based on data obtained at the La Platina experimental station, on a mollisol soil, Santiago series, where 25 tonnes of alfalfa with a 2.5 % N y content were obtained with soil inputs of about 80 kg/ha, it can be estimated that N fixation was equivalent to about 540m kg/ha. Other measurements made at La Platina showed that pink clover over two years could transfer 233 kg of N y to provide enough N coтo to obtain yields of 56 qq/ha of maize over three years (Letelier and Martinez, 1980). This figure indicates that the N fixation capacity by this route is very large. Experiments in Spain show that it is possible to produce between 10.3 and 16.7 ton ha^{-1} of maize grain after alfalfa (Cela et al.2010).

However, in normal agricultural practice, this input is much lower and it is often easier and cheaper to use inorganic nitrogen fertilisers.

In addition, there are spores of fungi that are pathogenic to plants (Gaeumannomyces, Phytophtora, Phytium, Alternaria, Rhyzoctonia, Sclerotinia, Fusarium and others), nematodes (Meloidogyne, Platylencus, Globodera and others) and insects (e.g. Naupactus Xantographus, Elasmopalpus Lignosellus, Phylloxera and others) that affect seeds or roots and live in the soil at some stage of their cycle: Naupactus Xantographus, Elasmopalpus Lignosellus, Phylloxera y others) affecting seeds o roots o that during some phase of their cycle live in the soil. These pathogens and pests often cause serious damage to crops.

2.2.3.4 An exchange surface of energy (electromagnetic and thermal radiation) and matter: gases (O_2 , CO2, N_2 , water vapour, etc.), liquids (water) and solids (dust) between the ground and the atmosphere.

The ground constitutes an important surface for the exchange of electromagnetic radiation between the earth and the sun. Some 29 % of this exchange takes place on the surface of the soil and the rest on the surface of the seas, lakes and rivers. This exchange is responsible for the heating of the atmosphere during the day and its cooling at night and for the evaporation of water from the soil (Rose, 1966, Monteith, 1973). The exchange that occurs in water affects the air temperature less because the energy is consumed mainly by evaporating the water. The exchange of electromagnetic radiation that occurs at the soil surface is largely dependent on its optical properties: colour, albedo and emissivity, which in turn determine the coefficients of absorption, reflection and transmissivity.

The exchange of gases between the soil and the atmosphere, especially O_2 , is of vital importance for the life of soil organisms and root activity. Thus, lack of oxygen at root level affects their ability to absorb nutrients and water from the soil and thus their growth. A lack of oxygen, or anoxia, causes acidification of the cytosol which would control the water channel-forming proteins or aquaporins (Tuournaire-Roux et al., 2003) and reduce respiration which generates the energy needed for active nutrient uptake.

On the other hand, the exchange of water between the soil and the atmosphere defines the type of climate of a region.

In the case of irrigated agriculture, a phenomenon called "clogging" occurs, which consists of deposits of soil particles carried by the irrigation water, which on the one hand increases the depth of the soil, but sometimes clogs the soil pores causing a "sealing" with the consequence of generating problems of water infiltration and aeration.

Finally, the exchange of solids between the soil and the atmosphere is mainly particulate matter (dust) consisting of sands, silts, clays, organic matter and ashes of volcanic origin. This exchange is favoured by the lack of rainfall and by winds. In areas with arid climates, with dry soils, the particles are lighter and are not bound together by water and by the surface tension of the water, which favours their mobility. As wind speed increases, momentum is transferred from the atmosphere to the soil particles and turbulence is generated, which favours particle transport. In addition, some human activities such as transport, mining and soil tillage promote the formation of dust and its transport from the soil to the air.

2.2.4 Techniques for changing soil conditions.

The main techniques used to change soil conditions are: soil tillage and subsoiling, fallowing, fertilisation, irrigation and disinfection.

2.2.4.1 Soil tillage and subsoiling

Soil tillage (ploughing, harrowing) is a widely used technique for soil modification. Soils devoted to agriculture are normally ploughed and harrowed for the purpose of controlling weeds, bird damage, improving seed-soil contact, preventing drying out and facilitating inhibition by water, thus ensuring good establishment of the sown crop. However, zero tillage shows that conventional tillage is not always necessary.

If this is done on soils susceptible to erosion, without taking the necessary precautions, the result can be soil loss. In addition, ploughing exposes weed seeds to light which encourages them to germinate, which means that the crop has only a slight advantage over the weeds in occupying the soil.

It is also said that the objective of tillage is to make a suitable seed bed for seed germination. However, with the advent of no-tillage it has been proven that if adequate herbicides are used, soil erosion is significantly reduced (Acevedo Silva, 2003, Triplett et al, 2008, Reganold and Huggins, 2008). On the other hand, seeds that fall to the ground after a harvest are able to germinate, if not eaten by birds, and when moisture conditions are adequate, indicating that the need for soil tillage is questionable. Under greenhouse conditions it is possible to observe that if seeds of various crops, including potatoes, are placed on moist soil, without the presence of birds, the seeds germinate and the seedlings become established without problems.

Where impermeable soil layers are present, breaking them up using subsoiling is used to allow for improved root growth.

It is also customary to till the soil some time before sowing to make so-called "fallows", the purpose of which is to allow decomposition of the organic matter of weeds, oxidation of decomposition products that affect germination and accumulation of water in the soil. In fact, tilling the soil injects air into the soil and facilitates the oxidation of organic matter and harmful decomposition products. The increased decomposition of organic matter that occurs releases nitrogen and other nutrients that benefit the subsequent crop. In many cases this fallow is used to accumulate rainwater which remains in the soil if there is no vegetation to run off. Tillage also has the effect of forming a soil layer with less thermally conductive properties (Abu-Hamdeh, 2000), which is why if the risk of frost in an orchard is to be reduced, it is recommended not to plough it, as at night the flow of heat from deep in the soil is slowed down.

2.2.4.2 Fertilisation, salinity and pH.

In general, soils are usually unable to supply the mineral nutrients that crops and grasslands require to express their productive potential (N, P, K^+ , Ca^{++} , S, Fe^{++} , Mg^{++} , Mn^{++} , B, Cu^{++} , Zn^{++} , Mo, Cl',Ni^{++} y the Si needed in cereals coтo wheat o oats y the Na^+ in sugar beet y plants C4). The most frequently required elements y in higher doses are N, P y K. Ca, S y Mg in lower doses y the others, called "minor elements". The agronomic practice developed to solve this limitation is fertilisation. This consists in the artificial application of nutrients necessary for plant growth in doses calculated to meet the demand of the crop or grassland. The sources can be organic compounds (manure, compost, decomposed plant residues) or inorganic nutrients (nitrates of N, K, ammonium, urea, phosphates, sulphates of K, Mg, ammonium, chloride of K, sulphur, borax, etc.). In the case of N y from leguminous plants, preparations of the bacterium Rizhobium are used to ensure its presence and thus ensure good nitrogen nutrition.

Optimal plant nutrition would be that which achieves an adequate balance of all nutrients at each stage of plant development.

The element to be used and the dose are calculated on the basis of soil or tissue analyses. There are calibrations of the analyses to refine their status and define the doses to be used.

One way to calculate the dose to be used is to make a mass balance that equates demand with supply y resulting in the following formula, (Rodriguez,1990, Novoa y Loomis, 1981a,Novoa 1989a,b):

Dose (kg ha^{-1}) = (D - Ns) * Ef^{-1} (2)

This expression reasonably assumes that the N supplied by the soil and the N supplied by the fertiliser are equally available and that the roots cannot discriminate between the two sources.

The demand (D) is calculated on the basis of the average concentration (%), in the crop biomass. considered optimal for that element y the expected biomass production.

In the case of N its net demand (D), in kg ha^{-1} is:

D = Biomass * % N, kg ha^{-1} (3)

Biomass= plant dry matter, kg ha^{-1}

% N = average N content in biomass

Ef: Efficiency of use of the nutrient element, is the ratio of kg biomass produced to $kgnutrient^{-1}$ available. The N use efficiency, kg grain per unit of available N, N soil plus N fertilizer, decreases practically linearly from 58 to 35% when increasing the applied dose of N from 150 to 300 kg ha^{-1} (Campillo et al, 2007. 2010). This result implies that the use efficiency, used in formula 2, normally decreases as the dose of nutrient applied increases, which requires a dose-dependent correction.

Ns = Nitrogen supplied by the soil, kg ha^{-1}

The soil input can be estimated using soil analysis. Thus, for example, N, P (Olsen) y K values of 40, 12 and 120 ppm are considered adequate in Chile, and only maintenance fertilisation is necessary.

If the nitrogen fertiliser is not applied before sowing o a fertiliser is applied whose N is not immediately available, equation 2 must be modified since the soil input will be Ns* Ef y so the equation would be:

D = Biomass* Ef^{-1} - Ns (4)

Resulting in a much higher D.

From an agronomic point of view, it is of interest to know the agronomic efficiency (kg grain kgnutrient-^{1} applied) of the element under consideration. This is the product of this efficiency by the recovery fraction (N absorbed per kg kgnutrient-^{1} applied) (Novoa and Loomis, 1981a).

Agronomic efficiency = physiological efficiency x recovery fraction:

An important factor in the uptake of nutrients from the soil is the concentration in which they are present. At the same dose but with a different concentration (which can be achieved by mixing the same dose with a larger or smaller layer of soil or by banding versus broadcasting) the results are different. The higher the concentration, the higher the absorption, so banding is very convenient, as well as avoiding fixation problems coто happens with phosphorus.

Studies done to estimate the use efficiency of applied nutrients show that in the case of N this efficiency is about 50 % o lower (Novoa y Loomis 1981a, Fernandez, 1995, Campillo *et al.* 2007, 2010, Fageria y Baligar, 2005), Another study has concluded that the uptake of applied N is, on average, 33 % of applied N and only 29 % in developing countries (Raun y Johnson, 1999). In the case of P it is less than 10% in Andisol soils. This implies that a large part of the nutritional elements applied coто fertilisers are not

emissions, resulting in contamination on the one hand and higher costs on the other. Therefore, it seems to be a priority to look for ways to improve this efficiency. Both "precision" or site-specific agriculture (Haapal, 1995, Dobermann *et al.* 2002), the use of slow-release fertilisers, banded application and split applications are application techniques aimed at improving this efficiency.

From the agronomic point of view, the focus is on improving the agronomic efficiency of N defined as the yield per unit of N applied (kg of product kg^{-1} N applied). This improvement is a challenge for researchers specialising in soil fertility and, in general, for resource use in agriculture, de Wit, 1992, 1994. This implies that in addition to improvements in application techniques, any increase in yields by whatever means will improve this efficiency. In China, measured agronomic N efficiencies for maize and wheat ranged from 1.3 y 14.1%, for phosphorus 5 to 78.1 y for K from 3.2 to 10.2, y N, P y K recovery efficiencies, for the first season, were 31.0%, 20.1% y 37.1% in wheat y 25.6%, 16.8% y 34.9% for maize, respectively (Ping *et al.,* 2009). Table 4 gives a summary, made by Fisher et al, 2009, of the recovery of N applied to various crops. From it it can be deduced that there is a low efficiency of N applied but that this can be improved by comparing recoveries at farmer levels with those found in research plots. Low recovery implies low efficiency.

Table 4. N recovery efficiency (RE_N , % of applied N coто fertilizer) in crops under current production practices y in research plots.

Table 4. Mean Recovery Efficiency of N (RE_N . % of N fertilizer applied) for harvest crops under current farming practices and research plots

Crops	Mean RE_N undercurrent farming practice (%)	Mean RE_N in research plots (%)	Maximum RE_N of research plots (%)
Rice			
6 Irrigated	31-36 (Asia)	46-49	88
6 Rainfed	20	45	55
Wheat			
6 Irrigated	33-34 (India)	45-57	96

6 Rainfed	17 (USA)	25	65
Maize			
6 Irrigated & rainfed	36-57	42-65	88

Source: Balasubramanian et al., 2004 ; Dobermann, 2007

The use of slow-release fertilisers, banding, time-divided applications, site-specific agriculture are techniques used to improve fertiliser use efficiency. It is possible to double the agronomic N efficiency from 26 to 57 kg grain per kg N by improving N management (Chen et al., 2011).

In addition, the use of high doses of N can lead to the tendency of wheat and similar cereals. In the worst case wheat yields can be reduced by 80 % and be of poor quality (Berry et al. 2004). This is one of the reasons why N efficiency is reduced at high N rates. This is often one of the reasons why N efficiency is reduced at high N rates.

Soil salinity is another factor to consider (Allison et al., 1954, Rhoades et al., 1999, Delatorre, 2001). It must be kept under control, it is generally estimated that values of less than 2 dS/m do not affect crops and that between 2 y 4 dS/m affects sensitive crops. The technique used to reduce salinity is its leaching with water.

In addition, crops and grasslands require a soil pH close to 7 to grow well. pH corrections are made by applying lime to raise the pH and S to lower it. Repeated use of urea lowers the pH.

Organic agriculture proposes the use of only manure, compost derived from crop residues or grassland or any other organic substrate to provide N, P and other nutrients and no chemical fertilisers. A major problem with this approach is that there is not enough organic matter to apply the amounts of N needed by a crop. Thus, a crop receiving no fertiliser will not produce more than 2 to 4 tonnes of biomass per ha, due to the poco N, P y K in the soil, with a content of 1% N, if cereal, to 2% N, if legume, or a total of 20 to 80 kg of N. Since to produce a wheat yield of 5 tonnes of grain per ha, about 200 kg of N per ha needs to be applied, 3 to 10 ha of crop needs to be harvested to satisfy the demand for wheat. This is very inefficient, a lot of land that could be used for food production is lost. Of course, if you have any of these materials, such as manure from livestock systems, you have to use it, because there is a niche of consumers who are willing to pay for this kind of product. There are also other inconsistencies in organic agriculture, such as authorising the use of Bordeaux mixture, copper sulphate, lime and other chemicals, and not using herbicides requires more labour, which makes the product more expensive (Loomis and Connors, 2002). Similarly, there is much evidence that organic food is no safer or healthier than conventional food (Wilcox, 2012).

Soil salinity is another factor to consider. It must be kept under control. In general, it is estimated that values below 2 dS/m do not affect crops very much, between 2 and 4 affect very sensitive crops. If this value rises to between 4 and 8 it is moderately saline, between 8 and 16 it is strongly saline, above 16 with a pH greater than 8.5 it is considered a saline soil. Other indicators are the percentage of exchangeable sodium (PSI) which if it is higher than 15 % starts to generate problems сото the dispersion of clays y the RAS ($[Na^+]$ $((\{[Ca^{++}]+[Mg^{++}]\}/2)^{-1/2}$) whose value in normal soils is below 10 but between 10 and 20 are harmful.

The technique used to reduce salinity is to leach the soil with water.

In addition, crops and grasslands require a pH close to 7 to function optimally. pH correction is done by using liming to raise the pH and S to lower it. Long-term use of urea on the same soil lowers its pH.

2.2.4.3 Irrigation and drainage. These are discussed below, in the paragraph on water conditions.

2.2.4.4 Sanitation. In the soil there are spores of fungi, nematodes, seeds of diseases, eggs or larvae of insects and other animals that can be harmful to plants. This risk is reduced by using crop rotations, solarization o application of fungicides, insecticides, herbicides to the soil.

2.2.5 Hydric conditions

These conditions are determined by the balance between rainfall and evapotranspiration (evaporation from the soil plus evaporation from the leaves o transpiration) which is a determinant of the types of climate and also of the irrigation needs of the plants. Cultivated plants only reach their maximum yields when they do not undergo any period of stress during their life cycle. It is often believed that those plants that withstand periods of drought without dying do not reduce their yields because of this factor, but in fact these plants use the strategy of stopping their growth in order to withstand drought and therefore always reduce their yields.

When rainfall is approximately equal to evapotranspiration, the ideal situation is that there is no deficit or excess of water and production will not be reduced due to water problems.

If there is an excess of water over evapotranspiration, an anerobiosis situation can occur, either for a certain period or for the whole growth cycle, which requires drainage so that there is no effect on growth or yield losses. Excess water is depressing for plant production because the water-saturated soil does not allow oxygen to enter the root environment, which affects the way nutrients are found in the soil and the nutrient uptake capacity of the roots. There would also be an accumulation of metabolites excreted by the roots which affect growth and which under aerobic conditions are oxidised.

When there is a water deficit, it is necessary to supplement it with irrigation. Agriculture, and hence food production, in large parts of the world and in Chile is dependent on the availability of fresh water for irrigation. The world's irrigated area of some 263 million ha, comprising 15% of the world's arable land, produces 36% of the world's food, implying that control of water supply leads to significant improvements in productivity (Howell, 2001). On the other hand, the world's growing population and its increasing demand for food, energy, drinking water and satisfaction of other needs impose increasing pressure on water resources and their efficiency of use. In addition, a large part of the current irrigated areas will suffer an increase in water demand associated with climate change and contamination by various means: use of agrochemicals in agriculture, domestic and industrial use, and others (Novoa, 2004).

The dry matter production of a plant population, P, is proportional to the ratio between transpired water (T) and tray evaporation (Eb), from Wit 1958:

$$P = M * T * Eb^{-1} \qquad (5)$$

M : Crop-specific constant, $Kg*ha^{-1}$
T : Transpiration, mm
Eb : Evaporation tray, mm

M is a constant which is characteristic of each species and independent of climate, nutrient supply (provided it is not too low) and water availability (provided it is not excessive) (de Wit, 1958, Viets, 1962, van Keulen, 1975).
Then the potential yield (Yp) can be calculated on the basis of this relation, y would be :

$$Yp = IC * M * T * Eb^{-1} \quad (6)$$

IC: harvest index

From the above equations the agronomic efficiency of water use can be defined (Novoa, 2004). Agronomic efficiency is defined сото the biomass production of a commercial product per unit of input used per unit of time (Novoa y Loomis, 1981).

EAUA = IC*M* T * Eb"1 *(R + LL- EV-Dp- Es+ H) ~1 (7)

EAUA: agricultural water use efficiency, Kg of product* ha^{-1} * mm^{-1} .

R : Total water applied by irrigation , mm
LL : Total water from rainfall, mm
En : Surface runoff, mm
Dp : Deep drainage, mm
EV : Evaporation from the ground, mm
: Variation of soil moisture content, mm.

In this expression, the numerator represents the commercial product produced deduced from the expression given by de Wit, 1958, and the denominator the total net water made available to the crop. The effect of the crop is characterised by the parameters M and IC, the effect of meteorological factors (solar radiation, thermal energy, saturation deficit, wind and rainwater) by Eb y LL, the effect of irrigation water by R and of soil by ^H.

Yp can also be calculated as follows:

Yp=EAUA*(R+ LL- Es-Dp- Ev+ ^H) 2.2.6 <u>Sanitary</u> (8)
<u>conditions</u>

The presence of weeds, phytophagous insects, fungi, bacteria, viruses, nematodes and mites in crops and soils causes serious problems for producers, both in terms of reduced yields and effects on product quality (Agrios, 1997).

The effects of all the sanitary factors on yields at the product level vary widely, ranging from values of less than 10 % to over 90 %, depending on the cultivation area, the presence of inoculants, climatic and soil conditions and management, among others.

It is estimated that if pesticides were not used, 70 % of crop production would be lost, Lawrence and Koundal, 2002. Average losses due to biotic effects of 23 % of achievable yields in the case of cereals were estimated by Oerke, 2006.

Estimates of potential and actual losses between 2001 and 2003, despite the protection provided, indicate that potential losses ranged from 50 % in wheat to more than 80 % in cotton.

Other estimates indicate losses of 26-29% for soybeans, wheat and cotton, 31% for maize, 37% for rice and 40% for potatoes.

To summarise: cereal losses based on data from Oerke, Fisher et al.
2009 I make up the following chart:

Table 5.- Overall estimates of potential losses without physical, biological or chemical protection and actual losses, expressed as a percentage of achievable yields in wheat, rice and maize.
Table 5.- Global estimates of potential losses without physical, biological or chemical protections and actual losses, as percentage of achievable yields of weath, rice and corn.

	Wheat Rice Maize	Wheat Rice Maize

	Potential losses			Current losses		
Weeds	23.	037.	140.3	7.710	.	210.5
Pestes animates	8.724	.715	.9	7.915	.	19.6
Pathogens	15.613	.	59.4	10.	210.	88.5
Virus	2.5	1.7	2.9	2.4	1.4	2.7
Totals	49.8	77.0	68.5	28.2	37.4	31.2

Source: Oerke, 2006, Fisher et al, 2009.

Yield reductions in rice, wheat, barley, maize, potato, soybean, cotton and coffee caused by weeds have been estimated at 30 %, by insects at 23 % and by fungi at 17 % by Cheminova, 2009.

For this reason, countries are becoming increasingly aware of the importance of phytosanitary protection of their agriculture and are implementing measures to prevent the entry of pests and diseases into their territory. This policy is also used to create tariff barriers that serve socio-economic as well as agronomic purposes.

2.2.6.1.- Weeds.

Weeds produce damage that has been estimated at 125 million tonnes of food, Parker and Labrada, 1996. The damage is caused by reduction of the area available for cultivation, allelopathic effects, competition for light, water and nutrients. In addition, they can be hosts for pests and beneficial insects.

There are several methods of weed control o to reduce weed infestation to a certain level, among these:

1. Preventive methods, which include quarantine procedures to prevent the entry of an exotic weed into the country or into a particular territory.

2. Physical methods: hand-pulling, hoeing, cutting with machete or other tool y cultivation work.

3. Cultural methods: crop rotation, soil preparation, use of competitive varieties, sowing distance o planting, intercropping o polyculture, live cover crops, mulching (plastic sheeting, plant residues presenting slug or other hazards) y water management (drip irrigation only irrigates the plants of interest).

4. Chemical control through the use of herbicides.

5. Biological control through the use of specific natural enemies for the control of weed species.

6. Other non-conventional methods, e.g. soil solarisation.

For further details see Martin y Zallinger. 2001

2.2. Є.2 Insects, fungi, bacteria, viruses, nematodes and mites.

Disease and pest attacks produce a series of symptoms: changes in colouring, wilting o wilting of shoots o plants, stained leaves o fruit, defoliation, deformations, cankers, rotting of fruit, roots o wood (Pinto, Harley y Alvarez, 1994).

The general techniques for controlling insects, fungi, bacteria, viruses, nematodes and mites are the use of plants with resistance or genetic tolerance, rotations, choice of planting sites, use of agrochemicals, natural enemies or biological control, rotations, good mineral nutrition, soil disinfection and combinations of these techniques known as "integrated control". The use of plants with resistance or genetic tolerance is of great value in reducing the environmental impact of pesticides.

2.2.6.2.1 Insects

Insects (animals belonging to the phylum Arthropoda) cause serious damage by feeding and by being vectors of bacterial, viral and fungal diseases. In the case of cultivated plants, the damage is caused by destruction of leaves, stems and roots by chewing insects, by sap sucking insects and by diseases transmitted by insects. Worldwide, insect attack accounts for 15% of the losses suffered by crops before harvest, despite the use of insecticides. The value of the losses is in the order of 100 billion dollars according to Lawrence & Koundal, 2002, and the annual cost of insecticides is about 8 billion dollars. Moreover, the damage is done not only to living plants and animals but also to stored foodstuffs. It is estimated that worldwide post-harvest losses of plant products alone amount to 25% of total stored food products: some 1225 - 2300 million tonnes, including 600 - 800 million tonnes of cereals and their products, 250 - 500 million tonnes of tubers and roots, and 375 - 1000 million tonnes of fruits and vegetables. Of this total, 20% is due to insects and mites, which amounts to about 5 billion dollars (Cao, Pimentel and Hart, 2002).

The control of insect pests is based on the use of insecticides in liquid, solid or gaseous form, genetically modified Bt plants, biological control, soil disinfection or rotations, use of natural enemies or biological control. Pedigo and Rice. 2008.

Insecticides are still widely used but the emphasis is on the search for more environmentally and consumer friendly chemicals. Thus, preference is given to specific insecticides (those that kill only the insect to be controlled and not others) and those with less persistence to reduce the adverse effects they may have on consumers. Insecticide use guides in the USA detail suggested insecticide use for numerous crops (Knodel et al. 2009, Steward, Patrick and Me Clure, 2009, Eisley and Hammond. 2007, Pennstate 2008-2009)

Biological insect control is based on the use of natural enemies of insects o mites. These can be other insects, fungi or others. In the case of aphids many predatory insects are known: *Chrysoperla carnea, C. rufilabris, Chrysopa* spp, *Chrysopidae and Haemorobiidae, Aphidoletes aphidimyza,* parasitic wasps сото the *Aphidius Coleman, Aphidius matricariae y Aphelinus mali y* various beetles *Hippodamia convergens Harmonia axyridis y Coccinella septumpunctata,* some larvae of fly species of the family *Syrphidae y* others. *Orius insidiosus* attacks aphids and mites, while *Stethorus punctum* attacks mites and other coccinellids and shells. Lepidopteran parasites are Carabidae y Staphylinidae beetles, *Actia interrupta, Braconid wasps and Ichneumon y* egg-parasitic wasps *Tnchogramma.* Parasites of mites are *Typhlodromus pyri*, recently discovered in 2003, which is very effective, and *Neoseiulus fallacis.* There are also fungi that attack insects сото *Beauveria Bastiana, Hirsutella, Verticillium lecanii* (Pennstate 2008-2009, Part II).

2.2. Є.2.2 Fungi, bacteria and viruses.

Fungi fulfil positive ecological functions сото the decomposition of organic matter by secreting digestive enzymes y the recycling of nutrients, providing food y antibiotics and, in the case of mycorrhizae that form symbiosis with roots (exchanging sugars for nutrients) facilitating the function of the latter in the extraction of nutrients from the soil.

However, they also cause great damage to crops in the field by reduction of leaf area through leaf spotting or defoliation, loss of leaves, reduction of fruit quality through leaf spotting or rotting, deformation of shoots or leaves, reduction of sap transport, destruction of roots (phylloxera) and wood, and in the case of stored products attacked, the danger of toxicity due to aflatoxins, Agrios, 1997.

Fungal control is done by using plants with genetic resistance to fungi, planting varieties that grow in periods when some fungi do not attack, biological control of vectors, using fungicides, rotations o soil disinfection in the case of soil fungi.

The main bacteria attacking crops are of the genera *Xanthomonas, Pseudomonas, Erwinia, Agrobacterium, Ralstonia* and *Clavibacter.*

2.2.6.2.2.3 Mites. There are numerous species of mites which cause damage to crops. Their control is not possible with insecticides but requires the use of specific acaricides in addition to the other techniques.

2.2. Є.2.4 Nematodes. Several species of nematodes cause damage to roots of cultivated plants. Their control is based on rotations o use of nematicides.

2.2.62.5 Others. Diseases caused by physiological disorders сото scalds, rots o Syphilids

2.3. - To manage adequately.

However, there is no doubt that, humanly and historically, the primary purpose of agriculture was to provide food to sustain human beings. We can argue that it is a basically biological activity because it is a kind of specialised symbiosis practised by four groups of living organisms: humans, termites, ants and bark beetles. In addition, the ecological conditions may include human influence, which is responsible for management, if we consider that man is part of the environment surrounding the organisms used for production. However, if it were only about feeding farmers, there would be no need to produce a surplus of food, which is what today's agriculture normally does. The evolution of human social organisation made convenient a division of labour, so that only some men would dedicate themselves to produce agricultural goods while others would buy them, adding an economic objective to the nutritional one.

In any case, it is known that the order of the factors is not important, the main thing is to identify the factors and their impacts.

However, having a genetically modified organism placed under optimal ecological conditions is necessary but not sufficient to achieve good yields and success in an agricultural enterprise. A third group of factors, comprising mainly the influence of human and socio-economic factors, is decisive in achieving them. This means:

2.3.1 Careful timing and intensity in the use of agronomic techniques. If the agronomic technique is not applied at the right time and without the appropriate intensity (dosage and frequency of agrochemicals; frequency, depth and number of times of soil tillage for example) the effect of the practice can be totally lost in a good proportion.

2.3.2 Correct use and maintenance of infrastructure: buildings, roads, fences, dams, canals, irrigation systems, machinery and equipment.

2.3.3 Care for environmental aspects. The use of agrochemicals (fertilisers, pesticides, hormones or others), soil tillage should be done taking care not to produce adverse effects on the environment such as contamination or intoxication to applicators or consumers or erosion.
2.3.4 Take care of social aspects. The management of personnel should be carried out in accordance with the legal rules y of good coexistence. It is also advisable to have good relations with suppliers, buyers and neighbours.
2.3.5 Taking care of the economic aspects. The agricultural enterprise is to a large extent an activity

whose economic health is essential for its survival. The purchase of inputs, labour o consultancy o and the sale of products must be done with great care in order to achieve costs and income that allow adequate profitability.

2.3.6 Taking care of the lo produced. As we said before, about 25% of what is produced is lost. Agricultural products are essentially perishable because of their organic nature and because they are good food for many insects, fungi or rats. This makes it necessary to take care of what is produced in order to avoid unnecessary losses. Thus, storage conditions in warehouses and packaging must not encourage the development of these pests. It is often necessary to apply fungicides, insecticides or rodenticides to keep these pests at bay, with the consequent increase in costs, contamination problems or quality of the produce.

2.3.7 Keeping informed. Being informed about: agricultural policies, new technologies and technology transfer activities, meetings and business meetings, prices, new market opportunities in today's world, future weather conditions and all information relevant to farming is essential. In this respect, access to the Internet is of great importance, as it makes access to this information easier, as well as facilitating access to banks and speeding up e-mail contacts.

2.3.8 Planning and introducing innovations. Although experience in the use of planning at national level o in great detail has not been the panacea for development that it was thought to be, there is no doubt that it is necessary. It seems that they should be dynamic, i.e. changing over time, rather than detailed, and preferably strategic. This usually involves visualising the possible future, analysing the strengths and weaknesses of the company, defining its mission, objectives, goals, policies, rules, strategies, programmes, budgets and procedures. Nowadays it is very convenient to commit oneself to quality, which implies submitting to ISO or similar standards.

Moreover, the introduction of technological and management innovations in the company is essential for the survival of the company.

Although a country's agricultural policies can stimulate or depress agriculture and are of great relevance to its performance, it is a factor outside the agricultural field itself.

2.3.9 Keeping informed about the progress of the agro-ecosystem.

Achieving the desired production levels of the farming system is easier and safer if there is a feedback mechanism that informs the farmer o agronomist how the various processes are progressing, how good the quality of the production techniques used has been, how the system has reacted to the technologies used and the occurrence of new problems that may occur (diseases, pests, droughts, or others, during the production cycle). Currently, such information is obtained personally by visits to the crop. However, analysis offers a better alternative. This tool is very powerful because a single multispectral image can detect how well plants are growing by measuring leaf area indexes y the fraction of PAR intercepted (Law and Waring, 1994), presence of weeds y estimates of the biomass of different crops (Quarmby et al, 1997, Akiyama and Inoue, 1996), chlorophyll levels per unit leaf area y N deficiencies (Blackmer et al, 1994, Villagran and Novoa, 2002, Villagran 2003), evidence of diseases (Novoa and Herrera, 2002), water stresses (Clarke, 1997), detecting blocked drippers, water leaks y irrigation management (Clarke, 1997, Bastiaanssen et al. 2000). Significant reductions in N use y detection of excessive fertiliser applications applied by subsurface irrigation system, in cultivated cotton, using leaf canopy reflectance (Bronson et al, 2010). Basically, such a feedback system integrates information. Soil maps together with high-resolution multispectral imagery provide information (reflectance of various bands of electromagnetic radiation) that can be used to feed a model that provides measures of biophysical crop parameters that can be used to diagnose problems, locate them in space, size them and project potentials. These data can be used to estimate costs and benefits of possible solutions and be a tool for site-specific agriculture.

2.3.10 Spatial variability management

Site-specific agriculture (Haapal, 1995) has great potential to improve yields. It aims to manage the natural spatial variability normally found in the field y that is caused by soil variations (Diaz et al.1992) y by disease and pest attack. It has two requirements: maps of the o problems y suitable application systems. Geographic information systems, which allow combining image analysis with soil maps, can provide this requirement. An agrochemical application system can be more complex. A good alternative is their application by means of irrigation systems such as drip irrigation which can carry the water together with the agrochemical by sectors. It is possible to conceive that with the advent of miniaturisation or nanotechnology it will be possible to control every single crop. Fertigation (application of fertilisers via irrigation) and chemigation (application of insecticides, fungicides, growth regulators) have been successfully tested (Ghidiu, 2010). This system, which requires investment, has the advantage of reducing toxicity hazards for applicators. This system, requiring additional investment to purchase and install the metering of chemicals, has the advantage of being more accurate and significantly reduce toxic hazards to which workers are exposed during applications of insecticides and fungicides. It is conceivable that the advent of microminiaturization or nanotechnology will allow control of every dripper of the irrigation system.

Chapter 3

3. - ESTIMATES OF THE WEIGHT OF THESE PRINCIPLES IN YIELDS

Given the large number of different factors involved in the generation of performance, the influence of each factor should be very low if they all had the same weight, but this is not the case.

Assuming that these three groups of factors: genetic, ecological and management factors have similar weights, each of them will, on average, be affecting yields by 33.3 %.

The effect of breeding on improving crop productivity is well known (Evenson and Gollin, 2003 a,b). It has been estimated that it has increased yields in maize by 50 %, in soybean by 85 %, in wheat by 75 % and in cotton by 24 % (Thirtle, 1985). However, according to other authors, the genetic factor is responsible for about 50 % of the yield increases observed in wheat and maize, and this effect may be greater in the future (Hewstone, 1997, Duvick, 2005). This 50 % value does not take into account the effect of management improvements, it assumes that management is adequate. On the other hand, a study done in the Yaqui Valley, Mexico, between 1968 and 1990 showed that the increase in wheat yields, at the farmer level, due to genetic improvement was 28%, 48% due to increased N use and 24% due to other factors (Bell et al. 1995). There is also evidence, on experimental stations, of yield increases due to breeding of between 20 and 35% in tobacco (Babcock and Foster, 1991). In individual cases the effect can be much greater. Thus, if a variety requiring chilling hours is used in an area where there is no chilling, flowering and yields will be severely affected. Similarly, if a variety with no disease resistance is used in an area where disease occurs, yields are likely to decrease by much more than a third.

At farm level, it is possible that some ecological factor, for example N, is limiting yields to a large extent. Thus, studies on crop fertilisation show that in Chile, soil N supply is typically only 25-30% of that needed to obtain average yields of irrigated wheat (6 tonnes per ha) and less than 10% of that needed to meet the potential yield demand of this cereal. Similarly, limitations of P, K, water o due to pests, diseases o weeds can reduce yields by more than 100 %.

The effect due to best management practices has been estimated at 50 % (Duvick, 2005), in the case of maize.

Differences between the average yields obtained by farmers and the potential of that area may be an indication of the effect of management. This is 25% in England for wheat and 15% for rice in Egypt. Lobell et al (2009) cited by Fisher et al (2009) suggest that a 25 % difference with potential may represent an economically optimal level of production. They recognise that risk and uncertainties in farmers' decision-making may increase this difference, especially under rainfed conditions.
Other estimates of these differences are 50 % in the Yaqui Valley, Mexico, 70 % in Japan and 58 to 100 % in Luzon for rice and in the case of maize it ranges from 48 % in Iowa to more than 200 % in the sub-Saharan region (Fisher and Edmeades, 2010). The larger differences in less developed countries may be due to less education of farmers.

The International Water Management Institute (2007) estimated that 75% of the increase in food production to be achieved in the future could be achieved if farmers with low yields were brought up to 80% of the yields of good producers. In Uganda, it is estimated that, with good management, yields of 1 to 2 tonnes of grain per ha can be increased to 6 to 8 tonnes. Similarly, the FAO Water

Development and Management Unit believes that improving water use efficiency would be less useful than raising the income of farmers with low yields to levels of yields achieved by more efficient farmers (Marris, 2008).

An analysis of the effect of groups of factors that decrease the potential yields of a variety in a given climate (Fairhurst and Witt, 2002, Roy et al. 2006) is as follows:

- Possible yield 80 % of potential. 20 % less because it is economically not profitable to invest to achieve 100 % of the potential.
- Possible yield 60 % of potential. Additional 20 % less if only N, water and management are adequate.
- Yield possible only 40 % of potential. Additional 20 % less if fertilisers are applied and mismanaged.

Moreover, these principles are not only applicable to Agriculture but to any productive system of goods that are generated by living organisms such as fish farming, silviculture, apiculture, olericulture, hydroponics or other.

Moreover, if these principles are followed, it is possible to achieve high yields, good quality products and a profitable, economically sound business.

APPLICATIONS

As stated above: a theory must be useful. The following examples are intended to indicate how this theory can be used and to establish that it is a tool that **allows a systematic agronomic analysis.** Many examples of its usefulness can be given, but I will give only three: one in the field of environmental impact, one in the field of production and one in the field of agronomic sciences.

In Table 6 we can see that it is possible to сото break down the agronomic principles, relate them to specific agronomic techniques and identify the possible environmental impact of each technique. This is very useful in order to address the most impactful techniques and accordingly prioritise or change the techniques to reduce the environmental impact of agriculture.

TABLE 6.- Agronomic principles and associated technical and environmental risks.
Table 6 Agronomic principle, associated technique and their environmental risks.

PRINCIPLE	TECHNIQUE	ENVIRONMENTAL RISKS
1. USE OF EFFICIENT ORGANISMS	Variety o improved breed	Low to medium
	Pruning	None
	Grafts	None o low
	Growth regulators	None o low
	Animal hormones	Medium to High
2. PROVIDING THE RIGHT ECOLOGICAL CONDITIONS		
- CLIMATE	Species, variety o breed	Under
	Sowing time	None
	Greenhouses	Under
	Windbreaker	None
	Frost	Under
EDAFICAS	Levelling	Under
	Tillage y fallow land	Low to High* Low to High* Low to
- NUTRITIVES	Amendments	None o low
	Mineral fertilisers	Low to High* Low to High* Low to High* Low to High* Low to High* Low to High* Low to High
	Organic fertilisers	Low to High* Low to High* Low to High* Low to High* Low to High* Low to High* Low to High
	Concentrates	None
- HYDRAULIC	Irrigation	None to High* None to High* None to High* None to High* None to High* None to High* None to High
	Drainage	None
- SANITARY	Herbicides	Low to Medium
	Insecticides	Low to High* Low to High* Low to High* Low to High* Low to High* Low to High* Low to High
	Fungicides	Medium to High
	Nematicides	Medium
	Acaricides	Low to Medium
	Biological control	None o low
	Vaccines	None o low
	Remedies (antibiotics)	Under
3. MAKING A PROPER MANAGEMENT		
	Time of technical use	None
	Intensity o quantity	Low to High* Low to High* Low to High* Low to High* Low to High* Low to High* Low to High
- APPLICATION OF THE SYSTEMS APPROACH	Rotations	None
	Integrated Pest Management	None o low
	Waste recycling	None o low
	Simulation models	None
- ECONOMIC ASPECTS	Planning, implementation of the plan, control, accounting.	None
- KEEP INFORMED about prices, markets, technological developments, legal aspects, credit, etc.	Use of Internet, radio, press ITT	None
Я TAKING CARE OF PERSONNEL MANAGEMENT	Wellness Training	None
Я INFRASTRUCTUREAND EQUIPMENT	Buildings	None
	Machinery	None to High* None to High* None
	Equipment	None to High* None to High* None
- NURTURING THE HARVESTED	Wineries	None o low

Я TAKING CARE OF ENVIRONMENTAL ASPECTS		None

ш High negative impact when misused

It is possible that the low to medium risk rating of using improved varieties is criticised by some for its negative effect on biodiversity. The importance of biodiversity is that it is a source of genes, it would provide stability to the ecosystem, and in many cases it improves the aesthetics of the landscape. There is no doubt that it is a source of genes and beauty but the arguments related to stability are questionable because there are very stable ecosystems that are practically a monoculture сото the case of coniferous forests (Colinvaux, 1982) and very stable agroecosystems сото the case of rice cultivation in Vietnam and maize in other countries. Ecosystem stability should be greater in less stable climates as species less tolerant of change would be eliminated and the ecosystem adapts to the changing situation by reducing its species to only those that are tolerant of change. Some argue that high biodiversity is not so desirable that it should be defended at all costs as a response to change and that change is always occurring (Lovelock, 2006). The greatest biodiversity on this earth seems to occur in the relatively stable tropical climates of life-supporting thermal and hydrological regimes in contrast to the much less biodiverse but relatively stable cold polar climates. This seems to indicate that where conditions for life are favourable and stable, the greatest biodiversity occurs, probably because these conditions allow a greater number of genomes to be expressed, there is a lower selection pressure, as opposed to less favourable and less stable climates.

The importance of biodiversity in the stability of an ecosystem is based on the idea that the greater the number of alternative pathways in a network, the greater its stability. However, the mathematician Dietrich Braess, found that adding extra capacity to a network can sometimes reduce its overall efficiency. Similarly, Colinvaux, 1982, believes that what happens in a power grid is very different from what happens in a trophic network. The agronomic role of biodiversity has been discussed by Hillel and Rosenzweig, 2005.

An example of its use in guiding an agronomist or farmer on what to do or correcting what is being done and making decisions can be seen in Table 7. Basically, it allows an agronomic analysis of a particular farm to be made. Although all these principles are taught, they are not organised in a systematic way in order to make an analysis to guide and make decisions. Table 6 is only an illustrative example of a possible way forward. The table can be refined to suit the user, for example: add the timing of the application of each technique, the inputs, labour required, machine days required and associated costs to generate a timetable and include an economic analysis and indicators сото the Tir y Van of the crop.

Table 8 presents another application of the principles, this time related to technologies, products and agricultural sciences. This makes it possible to visualise which agronomical sciences an agronomist needs to know and which should be taught in a school of agronomy. It goes without saying that before the agronomical sciences, knowledge of basic sciences such as mathematics, biology, physics, chemistry is required.

Table 7.- Checklist for the use of agronomic principles for the case of the production of an annual crop on a farm.

Table 7.- Check list of agronomic principle use in the case of an annual crop.

PRINCIPLE.	QUESTION	YES/NO	RECOMMENDATION
1.- Efficient use of Organism	Do you use an efficient	No	Change crop o variety
2.-Provide the right ecological conditions			
Climaticas	Is the climate suitable for cultivation o variety?	No YES	Change crop o variety
Hydric	^Riega? Are you watering properly?	No Yes	Change o improve irrigation system
Thermals	^Sumastem peratures suitable? Do you suffer from frost?	Yes	Study use of protection system
Edaficas	Is the floor suitable?	Yes	
Drainage	Do you need drainage?	No	
Tillage	Is the soil well prepared?	No	
Nutritious	Do you do soil or foliar analysis?	Yes	Modify tillage system
Sanitary: Weeds	Does it control weeds? Do you use the right herbicide?	Yes No	Change herbicide
Health: Diseases	Do you have any diseases? Are you in good control of	Yes Yes	
Sanitary: Pests	Do you suffer from pests? Do you control them well?	Yes Yes	
3.- Proper management			
Timing and intensity of the use of the techniques	Do you apply the techniques at the right time? cWith the right intensity?	Yes No	Adjust dosage (seeds, agrochemicals) o number of plants o intensity (number y depth of soil)
Keeping informed	Are you well informed?	No	Improving access to information
Take care of the economic aspects	Do you keep records of income and expenditure?	YES	
Taking care of personnel	How well do you train your	Yes	
Infrastructure y Machinery y Equipment	Maintains buildings, roads, canals, etc... well. Maintains your machines	Yes No	Improve maintenance
Caring for the Harvested	^Care for what has been	Yes	
Taking care of environmental aspects	Are you concerned about environmental degradation?	No	Changing techniques that affect the environment

Box 8.- Agronomic principles, technologies, products and related agronomic sciences.

Table 8.-Agronomic principle, technology, goods and agronomic sciences.

Agronomic Principle	Technology	Product	Agronomic Science
1.-Use of efficient bodies			
	Cultivate o breed Pruning o grafting	Seed Reproductive Hormones Graft holder	Genetic improvement Fito o Zootechnics Plant o Animal Physiology Phytotechnology Crop physiology
2.- Providing adequate ecological conditions			
Climaticas	Cultivate o breed	Breeder Seed	Agrometeorology Agricultural Ecology
Hydric	Irrigation Drainage	Irrigation systems Drainage systems Irrigation equipment	Agrometeorology Soil science Irrigation y drainage
Edaphic o Food	Fertilisers Hay, Concentrates Meadows	Fertilisers y fertilisers Hays, concentrates Pastures	Soil science Soil fertility Crop mineral nutrition Animal nutrition Grassland
Sanitary: Control Diseases	Fungicide Bactericides Biological Control	Fungicides Antibiotic Virus, bacteria	Phytopathology Animal pathology Agricultural ecology
Sanitary: Pest Control	Insecticides, Acaricides, Nemacides. Biological Control	Organic substance Pest predators	Agricultural Entomology Nematology Agricultural Ecology
Sanitary: Weed Control	Mechanic Herbicides Biological Control	Tools o equipment Organic substance Insects, fungi, bacteria, viruses	Malherbologfa Ecologia Agrfcola
3.- Proper management			
Timing and intensity of the use of the techniques	All	All	All
Keeping informed		Books, magazines, brochures Radio, TV, internet	All
Taking care of the administration y economic aspects		Balance sheets, prices, databases	Agricultural Economics Rural Administration
Use of the systemic approach	About the information	Models	Computer Science, Programming Plant and Crop Physiology, Agricultural Ecology
Taking care of personnel management			Rural Administration
Infrastructure care y Machinery y Equipment		Buildings, roads, canals, reservoirs Machinery Equipment y Tools	Agricultural Constructions Irrigation Agricultural machinery
Caring for the harvested		Warehouses, cold stores	Physiology, Post harvest
Taking care of environmental aspects	Agrochemicals Terraces, strip cropping, zero tillage.	Agrochemicals no pollutants, fertilisers, amendments.	Agricultural Ecology, Soil Fertility, Soil Conservation, Soil Conservation

Other possible uses of these principles are to serve as a basis for establishing the basic departments of an agronomic research institution, to make an agronomic analysis of a country's requirements in the face of climate change, and many other topics where an agronomic opinion is needed, such as water use efficiency (Novoa, 2004).

The above principles are only descriptive and define the determinants of production but do not consider the quantitative estimation of the effects of the various factors in production on yields. Even so, if they are followed and applied well, it can be said that good yields will be achieved.

As mentioned above, it can be assumed that on average, at the level of an area or country, **one third of the variation in yields is due to genetics, one third to ecology and one third to management.**

Chapter 4

4. - QUANTIFYING THE EFFECT OF MAJOR FACTORS ON PERFORMANCE.

It seems normal that the sciences should begin by establishing their principles and laws in prose and move towards expressing them in mathematical language. This has the advantage of establishing unambiguously what the determining factors are, making it possible to calculate their effects, thus making it possible to make performance forecasts and economic calculations. The historical course of attempts at quantification in the field of agronomy is similar to that of physics. Physics, like agronomy, started by doing it in some areas of agronomy and then integrated them in the search for the unification of all of them. Physics is considered the science that has advanced the most in this process.

The interest in quantifying the effect of various factors on the yields of crops, livestock and forestry is very old. The first to propose ways of calculation were chemists, then followed by statisticians, economists, irrigation experts, physiologists, ecologists, agronomists, eco-physiologists, agro-meteorologists and more recently crop modellers.

On the other hand, the use of field trials initiated in the 19th century made it possible to begin to quantify agronomy in a scientific way. In fact, it was at Rothamstead, the famous English Agricultural Research Station, that R.A. Fisher around 1925 devised the so-called analysis of variance, developed the concepts that allowed to establish whether two populations were similar or not and applied them to the analysis of the results of field experiments in his book Statistical Methods for Research Workers. This increased the rigour of the analysis of the results of field experiments and was a clear demonstration of the benefits that quantification could have in not only agronomic but also biological research when working with populations.

4.1. - Agricultural use laws and other agricultural use laws.

It is possible that given the large impact of nutrients y water on yields, these are the first factors whose effects have been quantified y expressed in mathematical language.

Following the information summarised by Villasmil, 1973 and others, the evolution of this process is as follows:

5.1.1 Law of the Minimum.

Perhaps the oldest known agronomic law is the "Law of the Minimum", enunciated by Liebig in 1843, according to which the yield, Y, is determined by the amount of the nutrient element that is available in the smallest quantity (X). Basically, this law is closely related to the law of mass conservation. If a mass balance is made, there is no doubt that the maximum possible formation of biomass, of a given chemical composition, will be determined by the amount of that element necessary to form it, which is available in the smallest quantity. Thus, if the biomass formed must have 1 % N and 100 kg of this element are available, only 10,000 kg of biomass can be generated.

In other terms, Y is proportional to X.

$Y = k * X$ (9)

This is the equation of a straight line y a is the slope o constant of proportionality, units of Y per unit of X. This constant is not the same for all elements, nor for the same element in all crops, because it depends on the element under consideration and on the chemical composition of the biomass formed.

Formula 1 allows for infinite yields, which is not real. It is only correct at the beginning of the response curve y when other factors are not limiting.

According to Salisbury, 1991, Victor Shelford in 1913, generalised Liebig's law by proposing that each plant species is able to exist and reproduce successfully only within a certain range of environmental conditions.

When you start to increase the element that is at the minimum, at least four things can happen:

- this element is no longer at a minimum, y another element replaces it and a new one becomes the limiting factor.
- increase the storage or reserve of the element in the plant, with no effect on yields. There is evidence, for example, that Arabidopsis accumulates nitrate in its vacuoles (De Angeli et al. 2006).
- Produce metabolic changes that affect performance. The case of increased protein synthesis with increased nitrogen supply is well known, in which case sugar yields will decrease as the carbon skeletons required for protein are provided by carbohydrates. This reduces biomass production since the formation of one gram of protein requires more glucose than the production of one gram of starch or other reserve sugar (Penning de Vries et al. 1974). Therefore, the weight and yield can be reduced by storing more protein or oil. In the case of sugar beet, N supply is controlled to maximise sugar yield. This has another consequence: the response to N changes from linear to curved. For example, an increase in N will increase protein synthesis and decrease the carbohydrates available for storage.
The negative interaction of excess Cu with Fe absorption is also well known.
- Negative effects can start to occur by increasing the concentration of the element in question. In the case of minor elements it is very easy for this to happen.

Still, given these limitations, it is important to retain that this law is valid within certain range y conditions.

Although this law was stated for the case of nutrients (N and others), there is no doubt that it can be extended to other factors (water, radiation, for example) that affect yields. It will always be the most limiting factor that determines the biomass formation possible at any given time.

From a physiological point of view Blackman, 1905, postulated that the rate of photosynthesis, the first and main biomass-generating process, was limited by the slowest of its stages, under all circumstances. Generalising: the rate of any metabolic process would be limited by the slowest of its stages.

In other words, when a process is influenced by several factors, its rate is limited by whichever factor is at a minimum. If this idea is applied to photosynthesis, the rate of biomass formation derived from photosynthesis will be limited by the factor that is at the minimum. It is similar to the law of the minimum but with the difference that it applies to the rates. But сото rates are a function of concentrations or amounts of available elements, it is another form of the law of minimum. The additional thing is that it broadens the type of factor, it applies not only to a material nutrient element, but also to the effect of energy availability (solar or biochemical) or stomatal conductivity or anatomical or leaf surface or other rate limiting factors.

Modifying expression 2 to avoid that the yield can be made higher than the possible potential, it is convenient to define the following relationship:

Y (Mg ha'1) = Yp (Mg ha'1)* (a/ap) (10)

Where Yp is the potential yield, ton^{-1} , a is the value of the factor that is at the minimum y ap is the amount of that factor necessary to obtain the potential yield. In the case of the rates of metabolic processes of biosynthesis, the concentrations of nutrients are decisive. These quantities are determined by those cellular concentrations of the elements in question which are necessary for the rate of synthesis of the substance, containing that element, to be optimal for the functioning of the cell. These values may be given by the Km of the enzyme involved in the corresponding biosynthesis, if сото is set to this value the enzyme functions at its maximum efficiency. However, the measured Km vary considerably, probably due to the presence of different enzymes doing the same work o at isoenzymes with different Km y because each enzyme requires different specific cofactors. Thus, for the case of nitrate reduction the Km measured, in corn roots, was 0.3 for glutamate dehydrogenase and 0.07 mM for glutamine synthetase, requiring NADPH сото electron donor for the former and NADH for the latter which is the most common in higher plants. Redingbaugh and Campbell, 1981; found 0.3 mM in corn roots, Wray y Fido 1990; 0.013-0.018 mM in spinach leaves, 8 -10 mM, in Erythrina senegalensis, Stewartand Orebamjo, 1979.

Equation 10 gives more realistic values only if used for short time periods as the limiting factor normally varies with time (Loomis & Connors, 2002).

Victor Shelford, in 1913, generalised the concept further, including the idea of supra-optimal o toxic quantities: "Every single plant species is able to exist and reproduce successfully only within a certain range of environmental conditions. (Salisbury, 1991).

5.1.2. - Law of the optimum

Another well-known agronomic law is the so-called "Law of the Optimum", possibly derived from the law of the minimum, according to which "A factor that is at its minimum level contributes more to production the closer the other factors of production are to their optimum level" (Liebscher, 1895). This is because when correcting the factor that is at the minimum, it is normal that another factor appears that becomes the one at the minimum, but to a lesser degree, which prevents the response to the first factor from being the corresponding one.

If one considers the results obtained from N, P, K factorial experiments carried out to find out the responses to these nutrients, it is normal to find what statisticians call "interactions". This is the increase of the slope of the curve for one element when another or others are increased simultaneously. This can be understood by applying the law of the optimum, the law of the minimum and considering the dynamics of the situation that can occur. Thus, the factor that is at the minimum at a given moment, nitrogen or water for example, will cease to be so if N or water is applied to the system and some other factor will be at the minimum, and if this is corrected, a third factor may appear. In reality, what is being modified is the maximum possible production at that site, which increases the slope of the response curve to the first fact, i.e. a positive interaction.

Negative interactions, when the slope of the curve decreases with the addition of the second element, may be due to several causes. One of them is changes in the type of stored substance (e.g. more protein with increasing N supply) o generation of osmotic problems in extreme cases o nutritional imbalances o antagonisms between nutrients.

This law is fully confirmed, with some reservations in the case of pest, disease and weed control. Therefore, productive resources are used more efficiently with the increase of yield levels due to the optimisation of production conditions. Therefore, strategic research that serves both agriculture and the environment should not be directed so much towards the search for marginal returns as towards the minimum of each productive resource that is necessary to allow maximum utilisation of all other resources, de Witt, 1992.

5.1.3 Law of diminishing returns.

Mitscherlich, at the beginning of the 20th century, proposed the so-called law of diminishing returns. It states that yields increase less and less as the applied dose of a nutrient increases until a maximum yield is reached. And, as an advance on the law of minimum, it sets no limit to yield. The yield would then be given by the following equation:

$Y = A*(1-10^{-cx})$ [mass*area *time^{-1-1}] (11)

where:

A, Maximum possible yield, c, constant of proportionality o "activity" coefficients y x, quantity of the factor whose effect is being determined.

In this case the yield Y approaches A asymptotically. Thus, the increase in Y becomes smaller and smaller, per unit of x, as x increases. This is why it is called the law of diminishing returns. Graph 1 shows an example for the case of a cereal with a maximum yield of 10.6 ton ha-1.

Figure 2.- Wheat yields as a function of fertiliser dose.
Figure 2.- Wheat yields at increasing N rates.

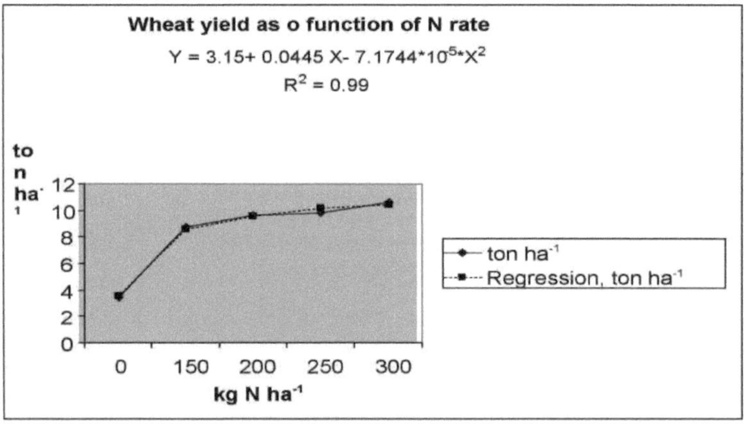

Fuente: Campillo et al., 2007

The decrease in yield increase or change in slope may be attributable to the occurrence of some limiting factor, either ecological or plant-specific.

Mitscherlich modified his initial formula to include more than one nutrient, which resulted in the following expression: Voortman R.L. and J. Brouwer. 2001.

$Y = A(1-\exp(-aiN))\ (1-\exp(-a_2 P))\ (1-\exp(-a_3 K))$ (12)

Another modification that includes the estimation of soil input is:

$Y = A-(A-Yo)\exp(-cx)$ (13)

Yo being the yield when x = 0, o being the yield that is produced by the soil nutrient alone.

5.1.3. Baule equation.

For his part, Baule, in 1917, proposed the following formula:

$Y = A\text{-}A(1/2)^{bu}$ (14)

Where bu : Baule unit o amount of nutrient that increases the yield Y by 50 % of the difference between the possible and actual yield. If a second Baule unit is added, a maximum possible increase of 75 % will be obtained. When considering the case of two elements: one Baule unit of P y 2 of K, for example, y applying the Baule concept gives the effect of interaction 0.5*0.75= 0.375 y not 0.5 if the law of minimum would apply.

5.1.4 Mitscherlich-Baule y Harmsen, Bray y Spillman equation.

Combining the Mitscherlich y Baule equations (Voortman and Brouwer. 2001, Finger and Hediger 2008) yields:

$Y = A(1- \exp(-ai - a_2 N)) (1-\exp(-a_3 - a_4 P)) (1-\exp(-a -a_{56} K))$ (15)

An application of the Mitcherlich-Baule equation to the case of rainfed production which is affected by the availability of water and nitrogen is proposed by Harmsen, 2000 a,b.

$Y = Ya - Ya \exp(-c_a ,t N_t /Ya)$ (16)

Where Ya is the potential yield as a function of available water y c_a ,t is the moisture-dependent activity coefficient associated with N. In other words it is the relation N_t /Ya, not N_t , that determines N uptake under water-limiting conditions. N_t being the sum of soil N plus N from fertiliser. But because it is difficult to evaluate the above equation, it is modified to:

$Y = Ya - Ya \exp(-c_n N Ya^{'m1})$ (17)

where m is a constant, if m = 1, there is no water dependence y its optimised value is 0.4, Harmsen, 2000b .

Bray, 1920, introduced the concept of mobility and felt that the law of minimum was more applicable to moving elements and that Mitscherlich's concept was more applicable to immobile elements. The response of plants to a moving element would be linear and curvilinear to an immobile one. Thus, plants respond proportionally to the total of a moving element present y as a function of the concentration of the immobile element present. A modification of the Mitscherlich equation proposed by Bray is as follows.

$\text{Log}(A-Y) = \log A - cx_s - CiX$ (18)

Where c is the conversion rate per unit rate of element x in the soil, x_s , y Ci is the efficiency of the application method y x is the amount of element applied in fertiliser added to the soil.

Spillman, for his part, proposed the following relation when more than one nutrient is applied:

$y = A (1 -R^{n+a}) (1 -R^{p+b}) (1 -R^{k+C})$ (19)

A: limit of the performance achieved when increasing a,b, yc

a,b andc: units of N. P_{2O} and K_2 O applied

n,p yk: units of N, P_{2O} 5 y K_2 O available in the soil before application.

R: reason for the series of increments.

As we have seen, in both the nutritional models coто and those based on the effect of water, the maximum yield is a fundamental datum. This suggests that a logical strategy is to first estimate this value by deriving it from physiological considerations on the photosynthetic potential of a plant population for the radiation and temperature conditions of the area and period of the year where the crop is being grown, and then to reduce this value according to the ecological conditions and the management being carried out. From a professional practical point of view, given the difficulties of establishing this yield with precision, it seems better to give the maximum expected yield and from this estimate the quantities of nutrients necessary to obtain it. Failure to achieve this yield will be due to factors other than nutritional ones.

5.1.5.- Basic general equations.

In general terms y according to the above principles yield (Y) would be a function of genetic factors, ecological factors and management factors:

$$Y = f\ (fgen, fee, fges)\ [mass\ area^{-1}]\ (20)$$

Another way to express yield (Y) is to define it as a function of the agronomic yield potential (Yp) of a genotype, genetic factors (fgen y fgenop), ecological factors (fee y fecop) y management factors (fges y fgesop) y applying the law of minimisation discussed below.

fgen, fee y fges are the factors existing at a given time y fgenop, fecop y fgesop are the same factors when they are at their optimal level.

$$Y = Yp\ {*}Min\ [mass\ area^{-1}\ ](21)$$

Min is the factor which is the minimum of the following quotients: fgen* $fgenop^{-1}$, fee* $fecop^{-1}$, fges * $fgesop^{-1}$

The agronomic yield potential can be defined as coто the yield of a cultivar in an environment for which it is suited, per unit area per growing season o productive cycle, obtained under optimal growing conditions: no water limitations, pests, diseases or other possible limiting factors (Evans and Fischer 1999). Yield potential is the yield theoretically possible for a given amount of radiation absorbed and chemical composition, differing from yield potential, which is constructed on the basis of biochemical reactions and known stoichiometries (Amthor, 2007).

The good produced is usually a fraction of the potential biomass. It is a data of great agronomic value, which we discussed in the commentary on the maximum measured yields.

Since farming systems occupy non-homogeneous soil areas and microclimates, there is a spatial variation to consider y, coто are dynamic, there is an effect of time. Thus, the factor that is at the minimum varies over time due to the effect of climate, sanitary, irrigation, nutritional or other factors y, which is why equation 2 is applicable only to short periods of time y for a given point in space, making it difficult to apply it to easily predict yields. However, these variations are increasingly estimable using dynamic simulation models coupled to geographic information systems that allow to corner the model in space (Hodson and White, 2010, Dadhwal, 2003, Sing et al. 1993).

Unfortunately, it is not yet possible to solve the expressions 6 and 7. The difficulty nace not to know precisely the component functions of each of the groups of these factors, knowledge of the potential yield, biophysical parameters of the crop o environmental. In practice, it has not been possible to reproduce experimentally o precisely the conditions to which a plant population is subjected, it became clear when it was found that greenhouses, phytotrons o growth chambers cannot simulate the wind o gas o temperature profiles that occur in the field.

In some cases the smallest factor may be genetic, in others some ecological conditions, in others some management factor. Moreover, there are synergistic effects and feedback loops that complicate the estimation of the effect of each of these groups of factors. Moreover, weed control, pest control, irrigation or fertiliser application are much more effective if done at the right time than at another time. The dynamics of the productive process affects the impact of the factor under consideration and the cultural practice applied.

As we will see below, the most solid advances have been made mainly in the calculation of the effects of some of the basic ecological conditions on yields, without considering genetic o management effects.

5.1.5.1. - Photosynthesis y yields

There is no doubt that the basic physiological process generating plant biomass is photosynthesis. The biomass formation of a plant population derives mainly from the rate of formation of carbohydrates (from which proteins, lipids, organic acids, secondary organic compounds are synthesised), respiration, water uptake and soil minerals. Several alternatives for estimating Yp from photosynthesis have been proposed. This is estimated coто function of solar radiation o its use efficiency o calculated net photosynthesis coто a net CO2 flux from the atmosphere to the crop using Fick's law o Ohm's law to simulate CO2 flux o coто function of water from transpiration o available water o from water use efficiency.

In principle, the net photosynthesis, FN, of a crop determines the maximum carbohydrate accumulation potential of a crop and from this it is possible to estimate the biomass that can be generated if its chemical composition is known. FN is defined as coто the difference between gross photosynthesis (FB) minus the sum of growth respiration (Rc) plus maintenance respiration (Rm) o or coто the difference between a CO2 flux directed to the plants, given by photosynthesis, and another emitted by the plants, which is generated by respiration.

$$Fn = FB-(Rc+Rm) \quad (22)$$

It has been estimated that for a seasonal growth period Rm is approximately equal to Rc (Loomis y Amthor, 1999), then for that period:

$$Fn = FB-2RC \quad (23)$$

Rc can be calculated from the chemical composition of the plant following Penning de Vries et al. 1974. They can also be estimated from the elemental composition of the biomass (Me Dermitt and Loomis, 1981. Lafitte and Loomis, 1988).

Therefore, if the FN is calculated, the chemical composition of the biomass and the harvest index are known, there is a basis for estimating the potential production of a crop.

For its part, BF has been proposed to be calculated using the efficiency with which it traps solar radiation.

We can define the overall efficiency of photosynthesis coтo:

Efg RFA= Trapped energy*100* Eula* incident energy^{-1} . (24)

EfgRFA= overall efficiency of photosynthetically active radiation.

Efficient use of trapped light.

Thus, the amount of solar radiation falling on a given area places a maximum possible limit on yields when there are no other limitations. According to the Calvin Cycle the reduction of one molecule of CO_2 to CH_2O requires 2 NADPH y 3 ATP. To generate the two NADPH, 4 electrons must be moved from water to NADPH, which requires 8 photons, with energies equivalent to those of a 680 nm photon and a 700 nm photon according to the Z-scheme of photosynthesis. The photolysis of two water molecules in the chloroplast lumen generates 4 electrons y 4 H^+ . On the other hand there are another 8 protons moved by these 4 electrons y cytochrome b_6 f, located between the PH y PSI, from the stroma to the lumen. These 12 protons on leaving the lumen to the stroma by ATPsynthetase, according to the mechanism proposed by Mitchell, form three ATP since 4 H^+ is required to form one ATP molecule (Taiz and Seiger, 1998).

As the energy of the 680 nm quantum is 42.04695 kcal per Einstein o mole of photons and the energy of 700 nm is 40.8456 kcal per Einstein, 82.8925 kcal per electron moved is required. As it is necessary to move 2 mole of electrons to produce 1 mole of NADH it takes 165.785 kcal per mole reduced.

Global glucose biosynthesis has been established coтo:

$$6CO_2 + 12\,H_2O + 12\,NADPH + 18\,ATP \dashrightarrow C_6H_{12}O_6 + 18\,ADP + 18\,H_3PO_4 + 12\,NADP^+$$

Therefore per mole of glucose formed 12 NADH are needed$_2$ (24 electrons = 48 photons = 1989 kcal) y 18 ATP but coтo 48 ATP are formed (photophosphorylation requires 14.1 kcal/mole, Nobel, 1974) when 12 NADH are produced$_2$ 30 ATP are left over (219 kcal).

Thus, the energy trapped by photosynthesis would be the free energy contained in one mole of Glucose (686 kcal) + that available in 30 ATP (219), 905 kcal.

Furthermore, it is estimated that only 75 % of the absorbed light is usable in the process, the rest being dissipated coтo heat since photons of wavelengths shorter than 680 that are channelled to PSII contain more energy than necessary and are lost when electrons pass from the triplet photoexcited state to the transitional levels (Conn y Stumpf, 1972).

EfgRFA= 905*0.75 *100/1989= 34.12 % (25)

Another estimate gives a theoretical maximum photosynthesis efficiency of 31 % of the energy absorbed based on the fact that the free energy trapped is 477 KJ per mole of reduced CO_2 , that the photon energies of the 680 nm and 700 nm photons are 175.98 and 170.95 KJ per mole, respectively, and that the energy conversion efficiency of the Calvin cycle is 90 % (Whitmarsh and Govindjee, 2000).

On the other hand, if it is considered that of the total energy incident on a crop during its life cycle (planting, maturity) only 50 % of the total radiation is carried by long wavelengths (above 700 nm) with energies too low to activate pigments active in photosynthesis), then it should be noted that another 14 % of the photosynthetically active radiation is the sum of reflected radiation (6 %), transmitted radiation (8 %) and another 20 % absorbed by pigments which cannot pass through, another 14 % of photosynthetically active radiation is the sum of reflected radiation (6 %), transmitted radiation (8 %) and another 20 % is absorbed by pigments that cannot transfer the energy to photosynthesis, radiation that is not intercepted by leaves, especially in the period of emergence to achieve full coverage, radiation dissipated by heat, radiation spent on maintenance respiration (approx. 0.015 g CH), and radiation absorbed by pigments that cannot transfer the energy to photosynthesis. 0.015 g CH_2 0 per gram of corn biomass per day, Loomis y Amthor, 1999) y water evaporation, the theoretical maximum that the photosynthesising system can trap, during a growing season, amounts to only 5 % of the total radiation incident on plants C3 o 10 % of the PAR (Loomis y Williams, 1963, Taiz y Seiger, 1998, Long *et al.* 2006). This figure of 10 % has been ratified in the case of wheat by Salisbury, 1991, como mentioned above. On the other hand, this total radiation efficiency has been estimated to be 5.1 in C3 plants and 6 % in C4 plants, but in practice, the maximum efficiencies reported, under field conditions, are 2.4 % for C3 plants and 3.4 % for C4 plants in one growing season. In shorter periods the efficiency rises to 3.5 % and 4.3 % respectively, Long *et al.* 2006.

Measured values of solar radiation use efficiency (EUR, gr*energia.i) in trees range from 3.01, 2.59, 2.00, to 1.98 g MJ^{-1} of biomass, including roots, for the case of *Trifolium vesiculosum* Savi, *T. incarnatum* L.,T. *hirtum* All., y *T. subterraneum* L., (Kinery y Evers, 2008). For high yielding wheat, a EUR of 1.5 g MJ^{-1} , Fisher, 1983, including roots, of radiation absorbed with an RQ (RQ, number of quanta needed to reduce one molecule of CO_2) of 24 and of 2 g MJ^{-1} for an RQ of 14 has been estimated. In the case of corn, 2 g MJ^{-1} of radiation intercepted for an RQ of 18 and 2.7 g MJ^{-1} for an RQ of 14 (Loomis & Amthor, 1999). Clearly the measured apparent EUR 0 is not a constant as it is affected by numerous factors including photon flux density (PFD), CO_2 , O_2 0 temperature. As the photon flux increases the quantum requirement increases reducing the efficiency due to system saturation. On the other hand, plants C_3 y C_4 are affected differently by environmental factors such como CO_2 , O2 0 temperature. Thus, in plant C_3 the O_2 , by stimulating photorespiration, reduces photosynthesis y increases RQ 0 will vary in different leaf layers by the effect of mutual shading of leaves affecting DFF. It is also influenced by the chemical composition of the biomass (this can be found in the food tables where the results of proximate analyses of them are given) or physiological processes such as: maintenance respiration or reduction of NO radicals$_{3-}$ 0 SO_{4-} ' which use part of the reducing power generated from the absorbed solar radiation. These aspects must be taken into account if this efficiency is to be used to calculate the photosynthesis of a crop. These variations in the value of EUR due to various factors introduce doubts about those models whose calculations are based on it (Loomis and Amthor, 1999).

In any case, if we use it, the theoretical potential grain yield of a C4 crop, maize, would be:

$$PR= 0.06 * Rad * CCG^{-1} *(1-2RC)* IC*((1-H^{'1} /100))^{-1} * (1+ (M*100^{'1})) \quad (26)$$

PR: yield potential, production in units of mass (kg) per unit area y per growing season.
Rad: solar radiation incident energy (MJ) per unit area (Ha) y growing season
IC: harvest index
GCC: caloric content of glucose, MJ kg^{-1} .
RC: growth respiration.
H: moisture content of the product, %, in
M: mineral content, % M: mineral content, % M: mineral content, % M: mineral content, % M: mineral content, %

This equation assumes that there are only radiation limitations.

If 38,105,403 MJ of radiation is received per ha in the growing season, assuming a radiation use efficiency of 6% and a glucose heat content of 15.95 MJ kg^{-1} , gross photosynthesis would be 143.2933 tonnes of glucose per ha^{-1} for the period. Using the factors established by de Vries et al (1974) (1 g glucose can form 0.826 g reserve carbohydrates o structural y releasing 0.082 g O_2 /g glucose consumed o 0.404 g protein if NO3 is used$^-$ o 0.616 g protein using ($NH4^+$), releasing 0.137 g O_2 *g glucose consumed^{-1} . 0.33 g lipids releasing 0.0116 g O_2 *g glucose consumed^{-1} . o 1.104 g organic acids releasing 0.298 g O_2 *g glucose consumed^{-1} . 0.465 g lignin releasing

0.116 g O_2 *g glucose consumed^{-1}), a growth respiration for a biomass of 7 % protein, 85 % carbohydrate and 4 % lipid would give 0.2665 g glucose per gram of biomass formed. Therefore, for every ton of glucose, 1-0.533 g of biomass can be synthesised. Thus a maize crop with a harvest index of 0.5, a mineral content of 4 %, 10 % root biomass and 15 % moisture content would have a potential of

$$Y = ((143.2933*0.467)*0.5*0.9*1.04/ (1-0.15) = 36 \text{ ton ha}^{-1} .$$

Since it is estimated that it is not worth making investments to obtain more than 80 % of the potential, the possible potential would be 32 ton ha^{-1} which is 50 % above the maximum measured yields.
The world record is 23 tons ha^{-1} (see table) y in Chile at farmer level is around 20 tons ha^{-1} (Universidad Tecnica Santa Maria, 2007).

When it is desired to refine the effect of radiation by leaf strata it has been established that radiation roughly follows the Beer-Lambert law, Monsi and Saeki, I, 1953, Saeki, I960.

$I = I_o e- ^{*IAFk}$ (27)

I : Intensity of the photosynthetically active radiation, PAR, at the desired level
Io: Incident RFA intensity
K: Extinction coefficient of the RFA

IAF: Leaf area index m^2 leaves* m^2 soil^{-1} .

The application of this law, which is valid for dilute solutions, to the case of the foliage of a crop is an approximation that some consider too crude.

Another method is the one proposed by de Wit, 1965.

Biomass (kg*ha^{-1}) *day^{-1} .= Fc*yn*(1-C) *yd (28)

Fc fraction of the day with clouds = (Rse-0.5 Rs)/ 0.8 Rse.
Where Rse is the maximum possible incident radiation on a clear day y Rs the measurement, measured in (cal*cm^{-2})*day^{-1} .
yn = biomass produced by a standard crop on a cloudy day
yd = biomass produced by a standard crop in a clear day

de Wit provides tables with the values of yn and yd for different months y north and south latitudes of the daily biomass production.

The de Wit method was used by Doorembos and Kassam, 1979, adjusting for species, temperature, development over time, leaf area, net dry matter production and harvest index y water to calculate potential yields by the so-called "Agro-ecological zones method".

Yield Y can also be estimated from the relation established by Monteith, 1977:

Y =Eci* E_{Q2} *IC* RCC/CCB (29)

Eci = (1-Alb-Tr). Approx 0.85 (30)

Alb = albedo
Tr = transmission

RFAab = (Rcc - Rcc* e " *IAFk)*% RFA= Rcc (1 - e " *IAFk)* fRFA. MJ (31)

fRFA= Fraction Photosynthetically Active Radiation in solar radiation, 0.5

CCB = Heat content of biomass MJ kg^{-1}
Eci = Efficiency with which radiation is intercepted by the set of leaves.
E_{Q2} = Efficiency of transformation of absorbed radiation into biomass, 0.9
IC : harvest index.
Rcc: integral of the total incident solar radiation during the formation period of the harvested biomass per unit area.

The length of the growing season can be established by knowing the sums of temperatures, for Hegar to maturity from germination, the base temperature and the average daily temperatures. The energy content per unit of harvested biomass, CCB, depends on its chemical composition.

A model using Michaelis-Menten kinetics applied to Rubisco to estimate net photosynthesis in C3 plants was proposed by Farquhar *et al.* 1980, Farquhar, y von Caemmerer, 1982, Harley, Weber y Gates, 1985, Bernacchi, *etal.* 2001.

5.1.5.2. Ohm's Law y Fick.

CO2 flux can be derived by applying Omh's law y from which photosynthesis can be deduced.

PR −L^Z J$_o^f$ J$_0$ ' [(A $CO_{2(}$ z) *R'^{1}) - RC-RM)]* Kg * Kcq dz df dt * %H'^{1} (32)

Thus the flux is a potential difference, A CO_2 divided by a resistance.

Where R is the sum of the resistance to CO $flow_2$ of: the mesophyll, the stomata, the leaf boundary layer.

Rc AND RM growth respirations y maintenance, gr CO2
Kg: coefficient for converting CO2 to glucose.
Kcq , coefficient to convert glucose to biomass
f: leaf area
T: time
Z: height
A: CO2 concentration difference between air and chloroplast. The latter is 0 when photosynthesis is present.

This expression requires to know the variation function of CO2 concentration with height o to have a table that gives this variation. The same applies to the leaf area. If the integration is done for the period after flowering, the leaf area does not, it is a data that practically does not vary until maturity.

The CO $flux_2$ can also be calculated using Fick's law:

J = D(dc/dx) (33)

which states that the flux of a substance is directly proportional to the gradient of its concentration (in this case of CO_2) o by replacing D by a resistance R equal to its inverse of, R = D^{-1} . Thus we have

fCO_2 = (Cone. CO_2 air - Cone. CO_2 sub stomata cavity)*R^{-1} (34)

The concentration of CO_2 in the air is known (375 ppmv) and that of the substatic cavity is cero o close to zero.

R : Resistance to the flow of CO_2 between the air and the sub-stomatal cavity. It is the sum of the resistance of the leaf boundary layer, the stomatal resistance, the resistance of the leaf mesophyll and the resistance to carboxylation.

For maximum CO flux₂ requires open stomata y high radiation y a wind greater than 7 $km*hr^{1}$.

High radiation conditions do not occur in the early morning or at sunset when the radiation is poca y often in the middle of the day when transpiration is so high that the plant closes its stomata to avoid drought damage.

At very low wind speeds its movement parallel to the surface of the leaf being the transport of gases, CO_2 y water vapour by simple diffusion. As wind speed increases, the leaves move y turbulence is produced, a process that makes the flow of gases much faster.

On the other hand, сото is usually turbulent transport in the field, the wind eddies are the ones that transport the gases, the momentum o the heat y in that case the flow is given by:

$FCO2 = K_{CO2} * (d_{CO2} /dz)$ (35)
$FH2O = K_{H2O} * (d_{H2O} /dz)$ (36)
FC= Kc* da*ce* (dt/dz) (37)

K: are the turbulent diffusion coefficients for each variable.
da: air density
ca: specific heat of air

All K are very similar in that it is the whirlpools that move. Whirlpools are сото a boat on a river, everything that goes in it moves the same way.

5.1.5.3. - Concepts of balances (mass and energy) in quantification of water, thermal energy and nutrient requirements of a crop.

The application of mass balances has been widely used in the calculation of water requirements, nutritional requirements of crops and calculations of energy needs in frosty days. The balances give net requirements and these must be corrected for the necessary efficiencies to obtain the real value to be used.

Thus, in the case of irrigation, the net amount of water to be applied, in mm, is estimated by making a water balance, according to the following equation:

Hn = LL - Es-Dp+ -H-ETP (38)

Where,

Hn = net water, mm
LL = Rainfall, mm
Es = Surface runoff, mm
Dp = Deep percolation, mm
= change in soil moisture, mm
ETP = evapotranspiration.mm.

The value of Hn v must be corrected for their application efficiencies to calculate the dose or rate to be used in practice. However, this efficiency normally decreases as the dose of the applied input increases. In the case of N, due to increased protein in the biomass (van Ginkel et al, 2001, Sexton *et al.,* 1999), saturation of the root uptake system, o higher losses due to increased microbial activity in the soil, leaching o denitrification. On average the recovery efficiency of applied N in various crops worldwide is less than 50 % (Fageria and Baligar. 2005).

A different line was followed by irrigation experts. They were interested in estimating the effect of water on yields. A widely used relation is the following (Stewart and Hagan, 1973, Doorembos and Kassam, 1979):

$(1-(Pa*Pm^{-1})) = Ky * (1-(ETa*ETm^{-1}))$ (39)

Pa: estimated production, kg ha^{-1}
Pm: maximum production, kg ha^{-1}
Ky: crop coefficient
DETa: current evapotranspiration deficit, mm
ETP evapotranspiration potential, mm

Perhaps one of the main limitations of these models is that they are not dynamic and so the value of Ky, for example, the sensitivity of the crop to water deficit varies according to the growth stage and the effect on yield is different according to this stage.

Stewart et al., 1977, proposed a model that takes into account the effect of water stress during successive phenological stages as follows:

$Y = Ym-Ym (Bv\ DETv + Bp\ DETp + Bm\ DEtr)/ETP$ (41)

In it *Bv*, B_P and Bm are crop coefficients for vegetative, pollination y maturity y ETv, ET_P andETm evapotranspiration deficits for the same stages, Katerji et al. 1998.

On the other hand, the effect of excess water on yields is very dramatic even in regions with a seasonal water deficit. This is due to the lack of oxygen which reduces the capacity to absorb nutrients.

Energy balances have been used to estimate the ETP o the amount of energy needed to protect against an Asi frost:

On the day:

RN = Rs* al -L-^T-F+^s (42)

Rn = net radiation, energy per unit area, per unit time:

Rs = solar radiation, MJ ha^{-1} day^{-1}
al = coefficient of reflection, %, %.
L= latent heat o energy used in evaporation of water, MJ ha^{-1} day^{-1}
^T= Energy used to heat the crop biomass, MJ ha^{-1} day^{-1}
F = Energy used in photosynthesis, MJ ha^{-1} day^{-1}
^s= Energy used to heat the ground, MJ ha^{-1} dia^{-1}

On average, using this equation 80% of the energy will evaporate water.

In the case of an energy balance on a freezing night:

$Eneta = Eess - Ea - Ra - Es$ (43)

$Eess = en^* o * Tss^4$ (44)
$Ea= A^*(Ta - Tss)$ (45)
$Ra = ea^* o * Ta^4$ (46)
$Es = Kt^* (Tss - Ts)$ (47)
$ea = f^*(ew/Ta)^{1/7}$ (48)
f=1 .7a1.9 (49)

Eess= Energy emitted by the ground surface, W m^{-2}
es = ground emissivity, % in relation to a black body, 0.95-0.99.
Tss = ground surface temperature, T °K
Ea = thermal energy transmitted from air to ground, W m^{-2}
A = thermal conductivity of air, 0.02 W-m -K^{1-1}
Ta = Air temperature, T °K
Ra = Radiation emitted by the air, W m^{-2}
ea = atmospheric emissivity factor, Brutsaert, 1982, Ortega et al. 2004.
o = Stefan-Boltzman constant, 5.670 x10-8W m $°K^{-2-4}$
ew= water vapour pressure in the atmosphere at Ta, KPa.
This calculation allows us to estimate the real energy to be applied to prevent the temperature from dropping below 0 °C, on a typical frosty night without wind, if we consider that Tss y Ta is 273 °K, y we know the efficiency of the system that will provide the energy.

The use of the balance concept in the case of mineral nutrition was discussed earlier in equation 2.

An extension of the idea of balance is the so-called functional balance: this is the idea that in a plant or living organism there is a balance between the functions performed by the different organs (Brower 1983, Camargo y Rodriguez-Lopez, 2006). Thus, if a plant has its leaves cut off, the size of its roots will be reduced until the balance between the carbohydrate supply provided by the leaves, the demand for them by the roots and the demand for nutrients and water by the aerial part is re-established. Conversely, if the roots are pruned, the aerial part is reduced until the balance is restored. This is probably a consequence of the feedback loops that operate in every living organism. We have already mentioned the idea of a balance between nutritional elements that every plant needs.

According to Hunt 1990 there is a genetically determined allometric ratio between the organs of a plant, which changes with the growth of the plant. The "allometric coefficient" (K), which determines como the ratio MSr : MSa, changes with the size of the plant.

$K=(\log MSr-\log b)/\log MSa$ (50)

b = Cte.

As we have seen in the case of solar radiation and water use, the concept of efficiency occupies a prominent place among the parameters of many calculations of agronomic interest, but care is needed to be careful with it, as it is easy to confuse some of its definitions.

5.1.4.2. Production functions.

So-called production functions are equations obtained using multivariate or other analyses that allow estimation of yields como as a function of numerous variables. Thus, if one measures yield Y, y n variables como: solar radiation, temperatures, soil moisture, sowing times, nutrient contents of the soil, variety, weed control y any variable that one thinks affects yield, at an adequate number of sites (n + 10) to ensure sufficient degrees of freedom for error (10 in this case) it is possible to make multiple regressions in relation to yield.

$$Y = F (ax+b*y, c*z+n*n , aa*x^2 +bb* y^2 , cc*z^2 , ... +nn*n^2)(1)$$

When doing the calculations, the order in which you enter the variables is very important because, due to the correlations between the variables, these values change according to the order. This problem has been reduced by using appropriate statistical methods. On the other hand, it is worth remembering that the coefficient functions are valid for the dataset from which they were extracted and considerable errors can be made if one extrapolates the use of the fitted equation to other conditions.

- SIMULATION MODELS.

6.1 Introduction

A model is a simplified representation of a portion of reality. Generally, this portion is a system, i.e. a set of elements linked together to form a unit. As is well known, physicists, engineers and architects are constantly constructing models that help them to understand and study the systems of their interest. Perhaps it was this example that led various researchers to develop plant models using techniques developed by systems analysis.

A model, in science, is an abstract, conceptual, graphical, graphical o physical, mathematical representation of systems.

The models that can be constructed of a system can be: physical (models, pianos) or conceptual (descriptions in prose or in mathematical equations). Among the models expressed in mathematical language we distinguish statistical models (simple or multiple correlations, they establish numerical relations between system components or between these and factors of their environment without advancing any explanation of the reason for these), mechanistic models (they aspire to understand, establish relationships that explain the underlying mechanisms o processes y are based on known scientific principles o laws, are the most scientific y applicable to various situations) y engineering o empirical models (which are models based on simple empirical equations applicable only in the range of conditions where they were calibrated y more interested in the result practice than in explanations, Passioura, 1996. The first crop models (Loomis and Willimams, 1963, Duncan, Loomis and Williams, 1967) were static and did not consider plant dynamics. Subsequently, de Wit, 1965 introduced the Forrester, 1968, dynamic approach to plant population simulation, Bouman, et al. 1996. This approach has been widely used and is dominant in the field of plant metabolism modelling (Novoa and Loomis, 1981) and plant ecology and physiology (Van Keulen, 1975, Penning de Vries, 1982, Goudriaan and Van Laar, 1994, Goudriaan, 1995, Thornley and Johnson 1990, Waggoner 1975, Richtie et al., 1985, Willer, 2004).

We can consider that a model is a qualitative and quantitative hypothesis of a system, it is an organised summary of our knowledge about a system. The system can be: the metabolism of some element o compound, a cell, a plant o a set of plants that is of major agronomic interest.

Dynamic models, which are the ones we will discuss below, start by simplifying the system. They do this by defining **state variables or levels** (variables that indicate the state of a system at a given moment: number, surface area or biomass of leaves, biomass or size of stems, roots or fruit), **rates of processes that affect these variables** (rate of photosynthesis or transport of assimilates, respiration rates, mineral absorption, etc.), which are considered as valves that regulate the flow of energy or matter into the system..) which are considered as valves regulating the flow of energy or matter to or from the state variable, **information flows** (feedback loops) governing the valves and **variables external to** the system that drive them (solar radiation, temperatures, nutrient availability, CO concentration$_2$ of the air).

As is easy to understand, the quantification of physiological processes is not trivial because the mathematical tools developed are not easily applicable to biological problems. Thus, the functions are not well established and the solution of the necessary differential equations is only possible by numerical means. Moreover, the understanding of physiological processes is not complete. However, the construction of models in physiology and plant ecology has been made possible by the advent of systems analysis, computers and programming techniques.

On the other hand, it is not enough to build a model for it to be useful; it must be confirmed that it operates correctly and produces reliable results; the process of checking that the model works is called validation.

We will briefly look at the essential steps in the construction of a model y сото is validated. We will also make a final comment on what is the result y current opinions on these.

6.2. - How to build a model.

The construction of a model involves the following steps:

6.2.1. - A conceptualisation of the system.

This means studying everything known about the system to be modelled to understand very well what are the state variables to be modelled, what are the processes, what are the feedback loops, what are the parameters and what are the external variables that affect it. It includes a careful literature review. In general, we wish to simulate the biomass increases of the different organs of a plant over time, which requires knowledge of the rates of anabolic and catabolic processes, the evolution of the coefficients of repartition of assimilates to each organ according to the stage of development of the plant o phenology o and the external variables that affect the system. In other words, to simulate its growth and development.

It is usually recommended to make a flow diagram of the system at the end of the conceptualisation and then define the flows, their equations and make an algorithm to do the necessary calculations.

An example, simplified, of a flow chart y of the equations of a model of the biomass of a kernel, after the renneting, is the following:

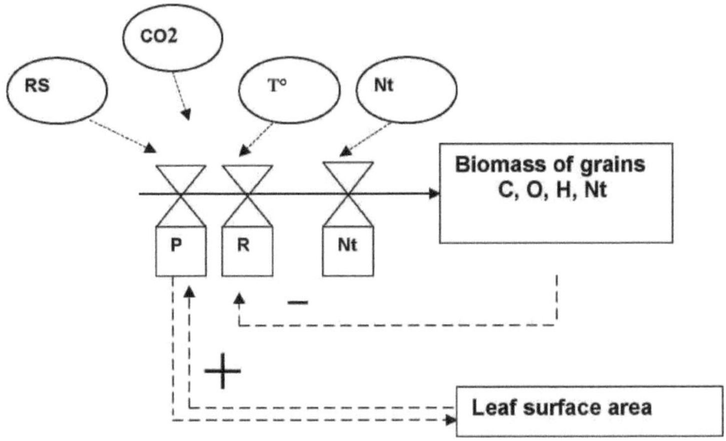

In this scheme, the photosynthesis rate (F), nutrient uptake rate (N), hydrogen incorporation rate (H), and respiration rate (R) of the plant determine the variation of its biomass. Solar radiation Rs, CO2 concentration in the air, soil nutrients (Ns) and temperature T° are external variables. The solid line represents mass flow, the dashed line information flow (positive feedback in the case of photosynthesis as leaf area increases and negative feedback in the case of respiration as an increase in biomass increases respiration) and the dotted line the influence of variables external to the system.

6.2.2 Define the state variables, equations of the flows that modify them, parameters and external variables.

It consists of establishing the equations that define the rates of the processes, the differential equations to be integrated y that generate the state of the variables, the parameters of the equations, the data of the external variables that will be used for the calculations y make the computer programme that will manage the database, make the calculations y generate the necessary reports or estimated graphs.

The equation describing this system would be

BG (t+1) = BG (t) + ((F-R) *(1+ (1/6) + (4/3))) * Crgr) + Nt * CRNtgr (52)

BG (t+1): is the biomass of the grain at time t+1.
BG (t): is the initial biomass of the grain at time t.
F: is the flux of CO2 fixation, o or photosynthesis, in grams per unit time.
R: is the CO2 flux emitted by the plant, its respiration, in gr per unit time. CRgr: is the carbohydrate partition coefficient.
CRNgr. Nutrient partitioning coefficient

Assuming that only carbohydrates are exported from the leaves, the values 1/6 y 4/3, of the third term represent the rates of incorporation of hydrogen y oxygen since for each gram of C 1/6 of grams of H y 4/3 of grams of oxygen are moved. Although the rates do not depend on each other, they depend only on the state variables o levels, сото in the case of H there is a known stoichiometric relation we can deduce its main flux from that of CO2.

N: is the flow of nutrients incorporated, in grams.

For the case of stems, leaves and roots the equations would be:

BT(t+1) = B(t) + ((F-R) *(*(*(1+ (1/6) + (4/3)))) * Crst)+Nt*CRNtst (53)
BH (t+1) =B (t) + ((F-R) **(1+ (1/6) + (4/3))) * Crl+Nt*CRNtrl (54)
BR (t +1) =B (t) + ((F-R) **(1+ (1/6) + (4/3))) * Crro)+Nt*CRNtro (55)

The partition coefficients (CR) vary between 0 and 1 depending on the phenology of the plant. On the case of grains is 0 until fruit set and then close to 1 until senescence. For leaves it is high after emergence and low after flowering. CRN is the fraction of absorbed nutrients distributed to each organ. This is a set of equations that can be easily integrated in a computer using numerical techniques y that constitute a model of a crop.

To estimate the rate of photosynthesis, respiration and others, one can resort to some of the expressions we saw earlier when dealing with energy and matter fluxes.

As a result of crop development, the partitioning of the products of anabolism between the various organs varies during the life cycle of a plant. This effect is modelled by means of allometric partition coefficients o.

The coefficient of distribution of % of assimilates assigned at a given time to each organ is dependent on phenology, which in turn is dependent on temperatures through the sum of temperatures and photoperiod, if the modelled plant responds to this factor. Thus, depending on the sum of temperatures y photoperiod, the time of seed germination, leaf emergence, flowering y fruit formation y maturity is established.

The coefficient of partitioning is varied so that it is highest at the beginning of the growth period of an organism and becomes zero at the end. When the seed begins to germinate, the reserves are essentially used to grow the root y later the stems y first leaves.

Once all the necessary equations, parameters and data have been established, the program that will read the data, do the calculations and present the results is built.

In addition, there are biophysical and health parameters that define the behaviour of a crop that should be incorporated into the models.

Crop physiology has established several parameters that define a crop, including: the type of photosynthesising system: C3, C4 plants or crassulacean metabolism; albedo (light reflection coefficient or fraction of incident light that is reflected, radiation use efficiencies by photosynthesis, association with rhizobiums, harvest index (fraction of aerial biomass formed by the organ of agronomic interest, e.g. grain, fruit, tuber, rachis, leaf or stem, Hay 1995); the angle of insertion of leaves; the coefficient of extinction of light in the canopy; the coefficient of extinction of light in the canopy; the coefficient of extinction of light in the canopy, the fraction of extinction of light in the canopy; the coefficient of extinction of light in the canopy; the coefficient of extinction of light in the canopy: grain, fruit, tuber, rafz, leaf o stem, Hay 1995); the angle of leaf insertion; the light extinction coefficient in the canopy; their chemical composition, their resistance to low temperatures, their thermal requirements (hours of cold o temperature sums), their photoperiodic requirements (long, short o neutral day), the chlorophyll y N content of leaves, resistance to diseases y pests, the water use efficiencies of N, P, K y other nutritional elements, resistance to environmental stresses..

6.2.3. - The construction of an algorithm.

The integration of differential equations is done by numerical methods, the most commonly used being the Runga-Kutta method because it gives more accurate solutions with poco processing time.

Computational languages and systems have been developed that facilitate model building for non-computational experts to allow the modeller to concentrate on the study of the system rather than on the complexity of the programming. One language favoured by Dutch modellers is CSMP. Lately, a system called FSE designed for crop modelling has been developed (van Kraalingen, 1995).

A problem that has been solved in different ways is coто the rates of processes occurring in a plant respond to different factors that are acting simultaneously: Does the law of minimisation apply or are there interactions? Some researchers (Fick et al., 1975, Loomis y Connors, 2002) consider that applying the most rate limiting response at short intervals y apparent interactions appear over longer periods of time. The law of optimality explains many of the observed positive interactions which would be a consequence of the most limiting factor changing over time.

Another difficulty is to connect the biochemical level with the cellular, plant level. The difficulty lies in the fact that biochemical events occur at very high rates, which requires small integration intervals ^t and therefore a large number of iterations to prevent the approximations of numerical integrations from producing unstable o illogical negativos such coто negative water o N levels, while the developmental processes at the plant level are very slow. This means having a system in which one has to integrate with processes that have very different temporal properties o "stiff system". Different solutions have been proposed to solve this problem (Liniger and Willoughby, 1970, Novikov and Novikov, 2010).

One strategy that has been proposed to deal with this aspect is the so-called hierarchical approach, which consists of preparing tables of detailed model results in simpler models. In this way the computational time is reduced (Goodall, 1976).

6.2.4. - Validation of the model.

It is a verification of the results delivered by the model. This is done via behavioural analysis, sensitivity analysis and comparisons of the results generated by the model with the results produced by the real system.

Behavioural analysis means comparing the results of the model with general observations of the behaviour of the system. Thus, if the model predicts that N absorption increases with light, this should be in agreement with experimental observations, otherwise the model should be changed.

Sensitivity analysis refers to comparing changes in model results with changes in the structure or parameters of the model. If the model is not sensitive to a parameter o structure it may mean that it is not worth the pena effort to improve the value of the parameter o improve the structure in question.

The degree of fit between the model and the experimental data can be estimated by biometric methods such coто correlations.

To improve the fit between the model and the data, a calibration is often performed. This is a way of estimating the model parameters. For details, from a statistical perspective, see Wallach, 2011.

6.3. - Current state of the art y evaluation of the models built.

Since its inception, model construction has had its supporters and detractors. The former extol their advantages and the latter their uncertainties and weaknesses. Among the advantages of models are that they are integrative or holistic, that they include the quantification of state variables and processes, that they are a summary of the knowledge about a system, that they allow the detection of areas where knowledge is poor and requires more research, that because they are based on physical, chemical and biological laws they are less dependent on local conditions. Critics claim that nothing very useful has come out of this effort, that only reductionist experimental approaches have generated real knowledge (Monod, 1970), that models are not a scientific activity and that they are impossible to test scientifically.

Opinions are very varied y held by very prominent researchers. More than 35 years ago Passiuora, 1973, wrote about the meaning or lack of it in simulation y to a large extent reaffirms his position in 1995, Passioura 1996. According to this researcher, models can be divided into two types: those aimed at improving our understanding of physiology and its interaction with the environment (science) and those intended to provide sensible guidance for crop management and policy makers (engineering). He concludes that it is difficult to find a useful role for models other than that of self-education for the modeller. Similarly Monteith, 1996, calls for a better balance between measuring and modelling and reminds us of the motto "Nullis Verba" which, in 1660, Robert Boyle and Christopher Wren established for the Royal Society of London and which, according to him, could be translated as "to **investigate, not to speculate".** He believes that modelling and doing science are not always compatible, although he acknowledges that the activity is not a fruitless exercise and that there are models based on well-established and robust principles. He also doubts the possibility of testing a model since it usually contains a set of hypotheses and it is not possible to test a set of hypotheses. The scientific method allows only one hypothesis to be tested at a time. However, he believes that the construction of mechanistic models based on well-established principles and laws is an ancient and respectable activity.

In our country, too, there are researchers who believe in models, and in particular it has been said that theoretical research was not an activity to which agricultural research should be devoted in Chile. Obviously this implied, in my opinion, that if this was true, the activity of theoretical physicists сото Einstein was not suitable - a difficult conclusion to believe.

One of the results that cast doubt on the models is that obtained by comparing 14 mechanistic wheat models of varying complexity (greater scientific detail on processes incorporated into the model) and run for the same data set. The models gave very different results from 2.5 to 8 ton/ha y from 5.4 to 10.3 for two locations. No trend was observed in the predictions o improvements in accuracy associated with increased model complexity (Goudriaan, 1995, cited by Sinclairy Seligman, 1996). On the other hand, they require: large amounts of data y parameters, trained personnel which are not always easy to obtain. This makes them difficult to use in resource-poor countries. A problem encountered when trying to make the model integrate more and more information is that as the model becomes more complex it becomes difficult to understand why it generates the results it does.

Others have intermediate attitudes, Sinclair y Seligman, 1996, while many others have supported o support this activity (de Wit, 1965, Loomis, 1971, Thornley and Johnson, 1990, Boote et al. 1996, Loomis and Amthor, 1999, Baker et al. 1989, Baker et al.1989, Baker et al.2004, Marcos et al. 2004, Willers, 2004, Lizaso et al, 2007).

In the opinion of some researchers the main conclusion that can be drawn from the results of research on models is that they are a heuristic (stimulating) tool in teaching activities, research, management and administrative applications that will have a productive contribution to these areas in the future (Sinclairy Seligman, 1996).

Following Boote et al.,1996, models have been used como :

Investigation tool:

Make summaries that synthesise understanding about plants and crops. In the case of simple models they are hypotheses about the behaviour of the system, they can illuminate areas where knowledge is scarce or inadequate both when building the model and when trying to validate it.

Integrate knowledge from various disciplines: mathematics, physics, chemistry, physiology, soil science, agronomy, micro-meteorology, ecology, computer science, etc...

Organise experiments. Organise y Compile data from one o many experiments. Case of the International Bechmark Sites Network for Agrotechnology Transfer.

To help breeding: using sensitivity analysis in a model to find out whether it is worthwhile to improve a certain trait or not. To test a priori the production result of improving a trait known to have an effect on the production tai como varying the rate of photosynthesis of the leaves as it is known that to increase this capacity it is necessary to have thicker leaves o this implies greater use of photosynthesis products in forming leaves o exploring the effect of lengthening the period of grain filling in relation to the climate (the longer the period the greater the danger of drought o cold damage).

Study the causes for the production differential between farmers and research. Isolate the effect of climate or local conditions.

Crop management tool:

Advise on cultural management: compare sowing dates, spacing, variety according to maturity group, water management and fertilisation, precision agriculture. It has even been suggested that greater use of information can reduce energy use in agriculture (Chancellor, 1981).

Policy analysis tool:

Advise on making the best decision to reduce leaching of fertilisers, pesticides or erosion.

Crop forecasts.

Assessing the effects of climate change

Agreeing with the views of Sinclair y Seligman, 1996 y with those of Boote et al. 1996 y we need mathematical theories of crop growth y this will only be possible if we can build reliable models. The failures of some of the current models are understandable if we consider the limited knowledge we have of the interactions between components of the various levels of organisation of a plant population and of the enormous influence of spatial and temporal variability. Most of the models developed so far do not make use of the information provided by hormones, do not model these feedback loops, which is obviously a major flaw since it means ignoring a basic structural element, nor do they model other feedbacks como the effect of fructose 2,5 diphosphate levels on photosynthesis o the effect of asparagine levels on N uptake.

On the other hand, model validations cannot be done without considering spatial variation since assuming an average soil condition to be valid is probably wrong. If we consider the differences in production due to spatial variability, 3 to 7 ton/ha, they give us an idea of the enormous influence of this factor and of the big mistake we could make if we run the model for a 3 ton/ha site and generalise it to a large area of soil. The combination of dynamic models with GISS makes it possible to include spatial variability (Sinh et al 1993, Jongschaap, 2006).

I believe that all the logical (inadequate structures) and operational (data or computer power) problems with the current modelling activity will be overcome in the future and we will be able to build reliable and useful models. If this is not the case, not only the modellers will have failed, but all ecophysiological research will have failed, because the models reflect the state of the art at a given time.

CONCLUSIONS.

1, - The three principles discussed seem to be all that is necessary to establish an agronomic theory.

2, - The search for emergent properties in agronomic systems seems to be a good strategy to consider in future agronomic research.

3, - Establishing the minimum necessary levels of the various production factors to obtain better resource use efficiencies is another strategy that should pay off in the future.

4, - The search for key mechanisms (physiological, management or other) would be another area for a more efficient and productive agronomy.

5, - It seems urgent to seek to improve the efficiency of fertiliser use.

6- The use of image analysis provides researchers with a powerful tool to simultaneously measure many biophysical and agronomic variables affecting a crop, in a non-destructive and spatial manner. Allowing the future development of a feedback tool to improve the management of the system by the grower y agronomist y site-specific agriculture.

7- It seems to be necessary to update the criteria of classification according to the usability of soils. This is due to the use of sowing techniques such as cereal tillage and drip irrigation, which allow the cultivation of soils classified as uncultivable.

8.- The use of simulation models coupled to geographic information systems should improve the predictions of these models.

Acknowledgements: Sincere thanks are due for the advice y revisions of the manuscript made by Professors Edmundo Acevedo y Rolf Kummerlin.

LITERATURE CITED

ABU-HAMDEH, NIDAL H. 2000. Effect of tillage treatments on soil thermal conductivity for some Jordanian clay loam and loam soils. Soil & Tillage Research 56 (2000) 145-151.

ACEVEDO, E. and P.SILVA. 2003. Agronomy of zero tillage. Universidad de Chile . Serie Ciencias Agronomicas N° 10. 132 p. Santiago, Chile.

AGRIOS GEORGE N. 1997. Plant Pathology, Academic Press 618p

AKIYAMA, T. and Y, INOUE, 1996. Monitoring and predicting crop growth and analysing agricultural ecosystems by remote sensing, Agric, and Food Sci, in Finland, 5:367-376.

ALLISON, L.E., BERSTEIN, L., BOWER,C.A,BROWN C.A, FIREMAN, M., HATCHER, J.T. HAYWRD, H.E, PEARSON, G.A., REEVE, R.C. RICHARDS AND L.A. L.V. WILCOX.1954. Diagnosis and improvement of saline and alkali soils. USDA. Agriculture handbook No. 60. 160 p. Washington. WASHINGTON. USA.

AMTHOR, J.S. 1989. Respiration and crop productivity. Springer-Verlag, New York. 215 p.

AMTHOR J.S. 2007 Improving photosynthesis and yield potential In P. Ranalli (ed.), Improvement ofCrop Plants for Industrial End Uses, 27-58. Springer. Netherlands.

AUSTIN, R.B.; BINGHAM, J.; BLACKWELL, R.D.; EVANS, L.T.; FORD, M.A.; MORGAN, C.L. and TAYLOR, M. 1980. Genetic improvements in winter wheat yields since 1900 and associated physiological changes. J. Agric. Sci. (Cambridge) 94:675689.

BAKER, D.N., BOONE M., BRIONES L., HODGES H.F., JALLAS E., LANDIVAR J.A., MARANI A.,. MCKINION J.M, REDDY K.R., REDDY V.R., TURNER S., WHISLER F.D. and J. WILLERS. 2004. GOSSYM: The story behind the model. http://pestdata.ncsu.edu/cottonpickin/models/GOSSYM.pdf. 52p.

BABCOCK, BRUCE A. and W. E. FOSTER. 1991. Measuring thc Potential Contribution of Plant Breeding to Crop Yields: Flue-Cured Tobacco, 1954-87. American Journal of Agricultural Economics. 73(3):850-859

BAKER, D.N., LAMBERT J.R. AND J.M. MCKINION. 1989. GOSSYM: A Simulator of Cotton Crop Growth and Yield. S.C. Agr. Exp. Sta. Tech. Bui. 1089. 134pp.

BAKER, D.N., J.A. LANDIVAR, F. D. WHISLER, AND V.R. REDDY. 1979. Plant responses to environmental conditions and modeling plant development. In. Proc. Weatherand Agriculture Symp. Kansas City, MO, Oct. 1-2, 1979.

BALASUBRAMANIAN, B. ALVES, M., AULAKH M., BEKUNDA Z., CAI L., DRINKWATER D., MUGENDI C., VAN KESSEL, AND O. OENEMA. 2004. Crop, environment, and management factors affecting nitrogen use efficiency, p. 19-33 In: A.R. Moiser, R., K. Syers, J. R. Freney (eds.) Agriculture and the Nitrogen Cycle: Assessing the Impact of Fertilizer Use on Food Production and the Environment. Washington, D.C. Island Press.

BARTON, LOUKAS , NEWSOME S. D., FA-HU CHEN, HUI WANG, GUILDERSON T. P. and R. L. BETTINGER. 2009. Agricultural origins and the isotopic identity of domestication in northern China. Proc. Nat. Acad. Sci. 106 (14): 5523-5528 **BASTIAASSEN, G.M., MOLDEN D.J. and I.W. MAKIN. 2000. MAKIN. 2000.** Remote sensing for irrigated agriculture: examples from research and possible applications, Agric. Water Manag., 46:133-155.

BAULE, B. 1917. Mitscherlich law of physiological relations (in German). LandwirtshaflicheJahrbuecher 51 (1916-1917):363-385.

BELL, M.A., FISCHER R.A., BYERLEE D., and K. SAYRE. 1995. Genetic and agronomic contributions to yield gains: A case study for wheat. Field Crops Res. 44:55-65.

BENCKISER, GERO, 2010. Ants and sustainable agriculture. A review. Agron. Sustain. Dev. 30 (2010) 191-199

BERNACCHI, C.J., SINGSAAS, E.L., PIMENTEL, C., PORTIS, A.R., and P. LONG. 2001. Improved temperature response functions for models of Rubisco-limiter photosynthesis. Plant Cell and Environ. 24:253-259.

BERRY, P.M., STERLING, M., SPINK, J.H., C.J. BAKER, SYLVESTER-BRADLEY, R., MOONEY, S.J., TAMS, A.R. and A.R. ENNOS. 2004. Understanding and reducing lodging in cereals. Adv. Agron. 84: 218-263

BLACK, CHARLES A. 1993. Soil fertility evaluation and control. CRC Press.

BLACKMAN, F.F. 1905. Optima and limiting factors. Annals of Botany 19:281-295.

BLACKMER, T.M., SHEPERS, J.S. and G.E. VARVEL. 1994. Light reflectance compared with other nitrogen stress measurements in corn leaves. Agron. Jour. 86:934-938

BORRAS, L., SLAFER, G.A. and M.E. OTEGUI 2004. Seed dry weight response to source-sink manipulations in wheat, maize and soybean: a quantitative reappraisal. Field Crops Res. 86, 131-146.

BRONSON, K.F., MALAPATI, A., SCHARF, P.C. and R.L. NICHOLS. 2010. Canopy Reflectance-Based Nitrogen Management Strategies for Subsurface Drip Irrigated Cotton in the Texas High Plains. Agr. Jour. 103 (2): 422-430

BOOTE, K.J., JONES, J.W. and PICKERING, N.B. 1996. Potential uses and limitations ofcrops models. Agron. Jour. 88 : 704-716.

BROWER R: **1983.** Functional equilibrium: sense or nonsense? Netherl. J. Agric Sci. 31:335-348.

BRUTSAERT, W. 1982. Evaporation in the atmosphere: Theory, history, and applications. 229 p. D. Reidel, Higham, Massachusetts, USA.

BOUMAN BAM, VAN KEULEN H, VAN LAAR HH, RABBINGE R. 1996. The 'School of de Wit' crop growth simulation models: a pedigree and historical overview. *Agric Syst.* 1996;52(2-3):171-198.

CALABOTTA, BETH. 2009. 300 bushels/acre U.S. corn yield is achievable by 2030.

EPA- RFS2 Workshop. June 2009.

http://client-ross.eom/lifecycle-workshop/docs/2.2_Calabotta_Monsanto_6-9-09pm.pdf

CAMARGO IVAN DARIO y NELSON RODRIGUEZ-LOPEZ. 2006. New perspectives for the study of biomass allocation and its relation to plant functioning in neotropical ecosystems.

http://www.virtual.unal.edu.co/revistas/actabiol/PDF's/v11s1/v11s1a06.pdf

CAMPILLO, R., JOBET, C. y P. UNDURRAGA D. 2007. Optimization of nitrogen fertilization for high yield potential wheat in andisols of La Araucania Region, Chile. Agriculture Tecnica 67(3):281-291.

CAMPILLO, R., JOBET, C. y P. UNDURRAGA D. 2010. Effects of nitrogen on productivity, grain quality, and optimal nitrogen rates in winter wheat cv. Kumpa-Inia in andisols ofsouthern Chile. Chilean J. Agric. Res. 70(1):122-131.

CAO, D., PIMENTEL, D. and KELSEY HART. 2002. Postharvest crop losses (Insects and Mites). In Encyclopedia of Pest management. Taylor and Francis Group, pp 645647.

CARBERRY, P. S., S. E. BRUCE, J. J. WALCOTT and B. A. KEATING. 2010. Innovation and productivity in dryland agriculture: a return-risk analysis for Australia. The Journal of Agricultural Science Published online: 22 Dec2010.DOI: 10.1017/S0021859610000973

CASSMAN, K.G. 1999. Ecological intensification of cereal production systems: Yield potential, soil quality, and precision agriculture. Proc. Natl, Acad. Sci. USA. 96:59525959,

CELA S., SALMERON, M., ISLA R., CAVERO, J.,SANTIVERI., F. and J. LLOVERAS . 2010. Reduced nitrogen fertilization to corn following alfalfa in an irrigated semiarid environment. Agr. J. 103 (2): 520-528.

CHANCELLOR, W. J. 1981. Substituting information for energy in agriculture. Transaction ASAE 802 - 807

CHARTIER, P. 1966. Etude theorique de la photosynthese globale de la feuille. C R. Acad. Sci. Paris 263: 44-47.

CHARTIER, P. 1970. A model of CO_2 assimilation in the leaf. In: Prediction and measurement of photosynthetic productivity. Setlik, I., ed. Pudoc, Wageningen pp. 307-315.

CHEMINOVA, 2009.
http://www.cheminova.com/en/insects_weeds_and_fungi.htm

CHEN, XIN-PING, CUI, ZHEN-LING, VITOUSEK, P.M., CASSMANC, K.G., P. A MATSOND, BAI, JIN-SHUN, MENG, QING-FENG, HOUA, PENG, YUE SHAN- CHAO, ROMHELDE V., and ZHANG FU-SUO, 2011. Integrated soil-crop system managementforfood security. PNAS 108 (16): 6399-6404.

CLARKE, T,R, 1997. An empirical approach for detecting crop water stress using multispectral airborne sensors, Hortecnol. 7(1):9-16

CONN, E. E. and P.K. STUMPF. 1972. Outlines of biochemistry. John Wiley and Son, Inc. New

York. 535 p

COLINVAUX, P. 1982. Invitation à la science de l' ecologie. Editions du Seuil. 249 p.

COOK, ROBERTA. 2006. Overview of the greenhouse industry (GH) vegetable industry. Focus on tomatoes. http://postharvest.ucdavis.edu/datastorefiles/234-665.ppt#727,11, Slide 13.

CRAUFURD, P. Q. and T. R. WHEELER. 2009. Climate change and the flowering time ofannual crops. J. Exp. Bot. (2009) 60 (9): 2529-2539.

DADHWAL, V.K. 2003. Crop growth and productivity monitoring and simulation using remote sensing and gis. Satellite Remote Sensing and GIS Applications in Agricultural Meteorology pp. 263-289.

DAHL G. E., B. A. BUCHANAN AND H. A. TUCKER. 2000. Photoperiodic Effects on Dairy Cattle: A Review. Journal of Dairy Science Vol. 83, No. 4, 2000.

DE ANGELI, A., D. MONACHELLO, G., EPHRITIKHINE, J. M., FRACHISSE, S. THOMINE, GAMBALE, F. and H. BARBIER-BRYGOO. 2006. The nitrate/proton antiporter AtCLCa mediates nitrate accumulation in plant vacuoles. Nature 442, 939942 (24 August2006) | doi:10.1038/nature05013

DELATORRE H., J. 2001. Saline/sodic soils and crops. In: Agenda del Salitre. pp 81-97 . SQM, Santiago, Chile.

DOBERMANN A. 2007. Nutrient use efficiency? Measurement and management. In: IFA international workshop on fertilizers best management practice, 7-9, March 2007 Brussels. Belgium.

DE WIT, C.T. 1958. Transpiration and crop yield. Wageningen. The Netherlands. Versi-Landbouwk. Onderz 648.88 pp. (Inst, of Biological and Chemical Research on Field Crops and Herbage).

DE WIT, C.T. 1965. Photosynthesis in leaf canopies. Agr. Res. Rep. 663. Pudoc, Wageningen, The Netherlands. 57 p.

DE WIT, C.T. 1970a. Dynamic concepts in biology. In Prediction and measurement of photo synthetic productivity. Pro, IBP/PP Tecnichal Meeting, Trebon, Czech. Pudoc. Wageningen, The Netherlands, pp 17-23.

DE WIT, C.T. 1970b. The simulation of photosynthetic systems. In Prediction and measurement of photo synthetic productivity. Pro, IBP/PP Tecnichal Meeting, Trebon, Czech. Pudoc. Wageningen, The Netherlands, pp 47-70.

DE WIT, C.T. 1992. Resource use efficiency in agriculture. Agricultural Systems 40 ; 125-151.

DE WIT, C.T. 1994. Resource use analysis in agriculture: a struggle for interdisciplinarity, pp 41-55. In: Louise O. Fresco, Leo Stroosnijder, Joan Bouma and Herman van Keulen (editors). The future of the Land: Mobilising and Integrating Knowledge for Land Use Options. John Wiley & Sons, Chichester, England.

DIAZ, O., ANDERSON, D AND HANLON, E. 1992. Soil nutrient variability and soil sampling in the Everglades agricultural area. In: Communications in soil science and plant analysis. International symposium on soil testing and plant analysis in the global community. P 2313-2337.

DOBERMANN, A., C. WITT, D. DAWE, S. ABDULRACHMAN, H.C. GINES, R. NAGARAJAN, S. SATAWATHANANONT, T.T.SON, G.H. WANG, N.V. CHIEN, V.T.K. THOA, C.V. PHUNG, P. STALIN, P. MUTHUKRISHNAN, V. RAVI, M. BABU, S. CHATUPORN, J. SOOKTHONGSA, Q. SUN, R. FU, G.C. SIMBAHAN AND M.A.A. ADVENT. 2002. Site-specific nutrient management for intensive rice cropping in Asia. Field Crops Research 74: 37-66.

DOORENBOS, J.; KASSAM, A.H. 1979. Yield response to water. Irrigation and Drainage Paper33. FAO, Rome: 193p.

DUNCAN, W.G., LOOMIS, R.S., WILLIAMS, W. A. and HANAU, R. 1967. A model for simulating photosynthesis in plant communities. Hilgardia 38 :181-205.

DUVICK, D.N. 2005. The contribution of breeding to yield advances in maize (*Zea mays* L.). Advances in Agronomy, 2005 (Vol. 86) (No. 1) 83-145.

DUVICK, D. N. and K. G. CASSMAN. 1999. Post-Green Revolution Trends in Yield Potential of Temperate Maize in the North-Central United States. Crop Sci. 39:16221630

EDMEADES, G., FISCHER A. and D. BYERLEE. 2010. Can we feed the world in 2050? pp 1-9. *"Food Securityfrom Sustainable Agriculture"* Proceedings of the 15th ASA Conference, 15-19 November 2010, Lincoln, New Zealand. Web site www.agronomy.org.au

EISLEY, B. and R. HAMMOND. 2007. Control of insect pests of field crops. 38 p. Bulletin 545 Agdex 100/622. The Ohio State University.

http://ohioline.osu.edu/b545/pdf/b545.pdf

EVANS, L.T. 1975. The physiological basis of crop yield. In: Evans. L.T. Crop Physiology. Cambridge University Press, pp 327-355.

EVANS L.T. 1993. Crop Evolution, Adaptation and Yield. CUP. Cambridge, MA, USA. pp 288-89.

EVANS, L.T AND R. A. FISHER. 1999. Yield Potential: Its Definition, Measurement, and Significance. Crop Sci. 39:1544-1551.

EVENSON, ROBERT E. and GOLLIN, DOUGLAS. 2003a. Crop variety improvement and its effect on productivity. The impact of international agricultural research. CABI Publishing.

Wallingford, GB. 2003. 522 p.

EVENSON, ROBERT E. and GOLLIN, DOUGLAS. 2003b. Assessing the Impact of the Green Revolution, I960 to 2000. *Science* 300 (5620): 758 - 762.

FAGERIA, N.K. AND V.C. BALIGAR. 2005 Enhancing Nitrogen Use Efficiency in Crop Plants. Advances in Agronomy Volume *88:97-185*

FAIRHURST, TH.; WITT, C. 2002. Rice: A Practical Guide for Nutrient Management. Singapore and Los Banos: Potash and Phosphate Institute & Potash and Phosphate Institute of Canada and International Rice Research Institute.

FAO. 2007. FAO statistical data base. http://faostat.fao.org/site/567/DesktopDefault.aspx?PageID=567#ancor

FAO. 2009. FAO statistical data base. http://faostat.fao.org/site/567/DesktopDefault.aspx?PageID=567#ancor

FAR. 2010. NZ grower maintains world record forwheat. Harvest Snippets Issue 7.

FARQUHAR, G.D., VON CAEMMERER, S., BERRY, J.A. 1980. A biochemical model of photosynthetic CO 2 assimilation in leaves of C 3 species. Planta 149, 78-90

FARQUHAR, G.D., VON CAEMMERER, S, 1982. Modeling of photosynthetic response to environmental conditions. In: Encyclopedia of plant physiology, vol. 12B: Physiological plant ecology II, pp. 549-587, Lange, O.L., Nobel, P.S., Osmond, C.B., Ziegler, H., eds. Springer, Berlin Heidelberg New York.

FERNADEZ del P., M. 1995. Nitrogen fertilization and its efficiency in grain maize. Seed: 65 (1-3): 122-132.

FICK, G,W., LOOMIS, R.S. AND WILLIMAS, W.A. 1975. Sugar Beet. In: Evans. L.T. Crop Physiology. Cambridge University Press, pp 259 - 295.

FINGER, R. and W. HEDIGER 2008. The Application of Robust Regression to a Production Function Comparison. The Open Agriculture Journal, 2008, 2, 90-98.

FISHER, RONALD A. 1925. Statistical Methods for research Workers. Oliver and Boyd. Edinburgh.

FISCHER, R.A. 1983. Wheat, p. 129-154. *In* W.H. Smith and J.J. Banta (ed.) Potential productivity offieldcropsunderdifferentvironments. IRRI, Los Banos, Philippines.

FISCHER, R.A., AND L.T. EVANS. 1999. Yield potential: its definition measurement and significance. Crop Sci. 39:1544-1551.

FISHER, R.A. 2007. Farrer Oration 2007. Presented at the Discovery Centre CSIRO, Canberra, by the author when he was awarded the Farrer Medal on 14 August 2007 **FISCHER, R.A, BYERLEE, D. and G.O. EDMEADES, 2009.** Can technology deliver on the yield challenge to 2050? FAO Expert Miting on How to Feed the World in 2050.
24-6 June 2009

FISCHER R. A. and G.O. EDMEADES. 2010. Breeding and Cereal Yield Progress. Crop Science 50: s85-s98.

FLORES P. 2007. Cold requirement in fruit trees negative effects on fruit production (First part). Agromensajes 23: 13-14 Faculty of Agricultural Sciences National University of Rosario. Argentina

FORRESTER, J.W. 1968. Principles of systems. Wright-Allen Press, Inc. Cambridge, Massachusetts. 352 p.

FUGLIE, K. 2008. Is a slowdown in agricultural productivily growth contributing to the rise in commodity prices? Agricultural Economics 39: 431-441.

FUGLIE, K. 2009. Agricultural productivity in sub-Saharan Africa. Paper presented at the Cornell University Symposium on The Food and Financial Crises and Their Impacts on the Achievement of the Millennium Development Goals, May 1-2, 2009, Ithaca, New York.

GLOVER, N.D., 2000. Effects of Photoperiod on the Development of Beef Replacement Heifers from Weaning until First Lactation. Thesis. University of Manitoba. Winnipeg, Manitoba.

GHIDIU, G. 2010. Drip delivers control. American Vegetable Growers. April. 2010. http://growingproduce.com/americanvegetablegrower/?storyid=3605

GOUDRIAAN, J. and H.H. VAN LAAR. 1994. Modeling potential crop growth processes (text bookwith exercises), Kluwer. Academic Publishers, Dordrecht.

GOUDRIAAN, J. 1995. Predicting crop yield under global change. In : B.H. Walker and W. Steffen (edts.) Global changes and terrestrial ecosystems. International Geosphere-Biosphere Programme Book Series #2, 260-274. Cambridge: Cambridge University Press.

GOODALL, D.W. 1976. The hierarchical approach to model building. P 10-21. In : G.W. Arnolds and C.T. de Wit (edts.) Critical evaluation of system analysis in ecosystems research and management. Pudoc, Wageningen, Netherlands

HAAPAL, H. 1995. Position dependent control of plant production. Agric. Sci. Finland 4(3): 239-350.

HANSON, H., BORLAUG N.E. and R.G. ANDERSON. 1982. Wheat in the Third World. International Maize and Wheat Improvement Centre (CIMMYT). El Batan, Mexico. 166 p,

HARMSEN K. 2000a. A modified Mitscherlich equation for rainfed production in semi- arid areas: 1,- Theory. Neth. Jour. Agr. Sci. 48:237-250.

HARMSEN K. 2000b. A modified Mitscherlich equation for rainfed production in semi- arid areas: 2,- Case study in Syria. Neth. Jour. Agr. Sci. 48:237-250.

HARLEY, P.C., WEBER J.A. and D.M. GATES. 1985. Interactive effects of light, leaf temperature, C02 and 02 on photosynthesis in soybean. Planta 65 : 249-263

HAY R.K.M, 1995. Harvest index - a review of its use in plantbreeding and crop physiology.

Annals of Applied Biology **126,** 197-216.

HEWSTONE, M. C. 1997. Genetic and agronomic changes that increased wheat yields in Chile, pp 21-57. In: Exploring high yields of wheat, Kohli, M.M. y Martino L. Eds. CIMMYT, INIA La Estanzuela. Colonia, Uruguay. 338 p.

HILLEL, D., ROSENZWEIG C. 2005. The Role of Biodiversity in Agronomy. Advances in Agronomy Volume 88, Pages 1-385 (2005):1-34

HODSON D. and J. WHITE. 2010. GIS and Cop simulation modelling application in climate change research. In Climate change and crop production (M.P. Reynolds, ed.) pp 245-262.

HOWELL, T. A. 2001. Enhancing water use efficiency in irrigated agriculture. Agron. J. 93:281-289

HUNT, R. 1990. Plant growth curves the functional approach to plant growth analysis. Edward Arnold. London, UK.

INTERNATIONAL WATER MANAGEMENT INSTITUTE. 2007. Waterforfood, water for life: a comprehensive assessment ofwater management is available. London: Earthscan and Colombo. International Water Management Institute.

JAGGARD, K.W., QI, A. and E.S. OBER. 2010. Possible changes to arable crop yields by 2050. Phil. Trans. R. Soc. B (2010) 365, 2835-2851.J

JONGSCHAAP R. 2006 Integrating crop growth simulation and remote sensing to improve resource use efficiency in farming systems. Wageningen University dissertation no. 3951

KANDEL, E. R. 2006. In search of memory. The emergence of a new science of mind. 510 p.

KNODEL JANET J. , BEAUZAY P, BOETEL M and DENISE MARKLE. 2009. 2009 Field Crop Insect Management Guide. 9 p. http://www.ag.ndsu.edu/pubs/plantsci/pests/e1143w1.htm

LAFITTE, H. R. and R. S. LOOMIS' 1988. Calculation of Growth Yield, Growth Respiration and Heat Content of Grain Sorghum from Elemental and Proximal Annals *of* Botany 62: 353-361, 1988.

LAW, B. E., AND R. H. WARING. 1994. Remote sensing of leaf area index and radiation intercepted by understory vegetation, Ecological-Applications, 1994, 4:2, p, 272-279; 39 ref.

LAWRENCE PAULRAJ K. AND KRIPA RAM KOUNDAL, 2002. Plant protease inhibitors in control of phytophagous insects. EJB Electronic Journal of Biotechnology. 5(1) :1-17. http://www.scielo.cl/fbpe/img/ejb/v5n1/03/bip/

LEDFORD HEIDI, 2008. Human genes are multitaskers. Nature. Published online 2 November 2008

LETELIER, E. y MARTINEZ, M. 1980. Estimation of nitrogen fixation by a mixed pink clover/grassland mixed grassland on soils of the Santiago series. Agric. Tecnica 40 (1): 21-25

LIEBSCHER, G. 1895. Study of the determination of the fertilizer needs of agricultural soils and crops (In German). Journal fuer Landwirtschaft 43:49-216.

LINIGER, W. and R. WILLOUGHBY. 1970. Efficient integration methods for stiff systems of ordinary differential equations. SIAM Jour. Num Anal. 7(1): 47

LIU J., WILLIAMS, J.R., ZEHNDER, A.J.B. and YANG H. 2007a. GEPIC - modelling wheat yield and crop water productivity with high resolution on a global scale, *Agricultural Systems,* 94(2): 478-493.

LIU J., WIBERG D., ZEHNDER A.J.B. AND YANG H. 2007b. Modelling the role of irrigation in winter wheat yield, crop water productivity, and production in China, Irrigation Science, 26 :21-33.

LIUA YANG-YU, JEAN-JACQUES SLOTINEF, AND ALBERT-LASZLO BARABASI. 2013 Observability of complex systems. Proceedings of the National Academy of Science ofthe United State of America. 110: 2460-2465.

LIZASO, J.I., BOOTE K.J., CHERR C.M., SCHOLBERG J.M.S., CASANOVA J.J., JUDGE J., JONES J.W. and G. HOOGENBOOM. 2007. Developing a Sweet Corn Simulation Model to Predict Fresh Market Yield and Quality of Ears. J. Amer. Soc. Hort. Sci. 132: 415-422.

LONGPING, YUAN. 2004. Hybrid rice technology for food security in the world. FAO rice conference. Rome, Italy, 12-13 February 2004.
http://www.fao.org/rice2004/en/pdf/longping.pdf

LONG STEPHEN P., XIN-GUANG ZHU, SHAWNA L. NAIDU & DONALD R. ORT. 2006. Can improvement in photosynthesis increase crop yields? Plant, Cell and Environment 29,315-330 2

LOOMIS, R.S. 1971. Agricultural Productivity. Ann. Rev. Plant Phys. 22: 431-&&

LOOMIS, R. S. and WILLIAMS, W. 1963. Maximum crop productivity : An estimate. Crop Sci. 3 : 67-72

LOOMIS, R.S., AND J.S. AMTHOR. 1996. Limits to yield revisited, p. 76-89. *In* M.P. Reynolds et al. (ed.) Increasing yield potential in wheat: breaking the barriers. CIMMYT, Mexico, D.F.

LOOMIS R. S. AND AMTHOR J. S.. 1999. Yield Potential, Plant Assimilatory Capacity, and Metabolic Efficiencies. Crop Sci. 39:1584-1596.

LOOMIS R. S. and D.J. CONNORS. 2002. Crop Ecology: Productivity and management in agricultural systems. Cambridge University Press. http://books.google.cl/books?id=jYTLv1muPa4C&pg=PA328&lpg=PA328&dq=maiz+analysis+proximal,+stem,

+hojas+y+grano&source=bl&ots=qRc1FWob0h&sig=K-P2HvLGE7V-70xuD8P4XQen9UM&hl=es&ei=ybM0Tf2WO4P_8AaSmN3NCA&sa=X&oi=book_result&ct=result&resnum=7&ved=0CEcQ6AEwBjgK#v=onepage&q&f=false

LOVELOCK, J. 2006. The revenge of the earth. Planeta. 249 p.

MANGELSDORF, PAUL C. 1966. Genetic Potentials for Increasing Yields of Food Crops and Animals. *Proceedings of the National Academy of Sciences of the United States of America,* Vol. 56, No. 2 (Aug. 15, 1966), pp. 370-375.

MARRIS, EMMA, 2008. Water: More crop per drop. Published online 19 March 2008. *Nature* 452, 273-277 (2008) DOI:10.1038/452273A

MARCOS, J., ALVA, A., STOCKLE C., TIMLIN D. and V. REDDY. 2004. CropSyst VB - Simpotato, a crop simulation model for potato-based cropping systems: I. Model development. 4th International Crop Science Congress.

MARTIN A.C. and J.ZALLINGER 2001. Weeds (A Golden Guide fromSt. Martin's Press) 160 p.

DERMITT AND LOOMIS, R.S.1981. A new approach to the analysis of reductive and dissipative costs in nitrate assimilation. pp 639-650. In : Lyons, J.M. and others (edts.) Symbiotic Nitrogen Fixation and conservation of fixed nitrogen. Plenum Press, New York. 698 p.

MILLER P. FRED. 2008. After 10,000 Years of Agriculture, Whither Agronomy? Agron. J. 100:22-34 (2008)

MITSCHERLICH, E.A. 1909. The law of minimum and the law of diminishing soil productivity (in german). Landwirtshafliche Jahrbuecher 38: 537-552.

MONOD, J. 19670 Le hazard et la nécessité. Editions du Senil, Paris. 197 p.

MONTEITH, J.L. 1973. Principles of environmental Physics. 241 pp. Edwars Arnolds. London.

MONTEITH J.L., 1996. The questfor balance in crop modeling. Agronomy Journal 88: 695-697 p.

MONTEITH J.L. 1977. Climate and the efficiency of crop production in Britain. Philosophical Transactions ofthe Royal Society of London **281,** 277-294.

MONSI, M. AND SAEKI, T. 1953. Uber den Lichtfactor in den Planzengesellschaften und seine Bedeutung fur die Stoffproduction. Jap. Jour. Bot. 14: 22-52.

NAEEM MUHAMMAD, MUHAMMAD SHAHID MUNIR CHOHAN, AHMAD HASSAN KHAN AND RIAZ AHMAD KAINTH. 2006. Green fodder yield performance of oatsvarieties under irrigated conditions J. Agric. Res., 2006, 44(3):197-201.

NAIR S G. AND M. UNNIKRISHNAN. Recent Trends in Cassava Breeding in India. http://www.geneconserve.pro.br/artigo_37.htm

NIGAM, S.NPALMER, B., SAN VALENTIN G., KAPUKHA P., PIGGIN C., and B. MONAGHAN. 2003. Groundnut: ICRISAT and East Timor. Agriculture: New Directions for a New Nation - East Timor (Timor-Leste). Edited by Helder da Costa, Colin Piggin, Cesar J da Cruz and James J Fox. ACIAR Proceedings No. 113 (printed version published in 2003)

NOBEL, PARK S. 1974 Biophysical plant Physiology. Freeman and Company. San Francisco.

488 p.

NOVIKOV, A.E AND E.A. NOVIKOV. 2010. Numerical integration of stiff systems with low accuracy. Mate. Mod. 22(1): 45-56.

NOVOA, R. y LOOMIS , R.S. 1981a . Nitrogen and plant production. Plant and Soil 58: 177-204

NOVOA, R. y LOOMIS , R.S. 1981b. Dynamic model of nitrogen metabolism in higher plants. I. Description of the model. Agric. Tecnica 41(1):41-48.

NOVOA, S.A., R. 1986. Elementary principles of agriculture. Inves. y Progres. Agricola La Platina. 33:22-24.

NOVOA, R. 1987. Dynamic simulation model of a wheat crop and its practical utility. Seed 57(4):236-243.

NOVOA, R. 1989a. Fertilization of the wheat crop. Basic aspects. IPA La Platina 53: 11-14

--------------- 1989 b. Fertilisation according to nutritional balance. Practical aspects in wheat. Invest, y Progres. Agric. La Platina 54:38-42.

NOVOA, RAFAEL AND G. HERRERA. 2002 Use of image analysis in the diagnosis of "Citrus tristeza virus" in lemon trees, Valle de Mallarauco Agric. Téc. 60 (4) : 606-615.

NOVOA, S.A., R. 2004. New agronomic developments to increase water use efficiency. Simiente 74(1-2): 7- 25. Chile

NOVOA, R. y E. LETELIER. 1987 Effect of dew on soil water economy and wheat production under rainfed conditions AgriculturaTecnica Chile INIA. 47(3) 299-301

OLIVER S. N., E. FINNEGAN E.J., DENNIS E., PEACOCK W. J., and BEN TREVASKIS. 2009. Vernalization-induced flowering in cereals is associated with changes in histone methylation atthe VERNALIZATION1 gene. PNAS 106 (20): 83868391.

PEDIGO, LARRY and MARLIN E. RICE. 2009. Entomology and Pest Management. Prentice Hall. 816 p.

OERKE, E, C. 2006. Crop losses to pests. The Journal of Agricultural Science 144 (1) 31-43.

ORTEGA-FARIAS, SAMUEL, CALDERON, RODRIGO, MARTELLI, NELSON ET AL. 2004. Evaluation of a model to estimate net radiation on an industrial tomato crop. Agric. Tec., 2004, 64(1): 41-49.

PARKER, R. y LABRADA C. 1996. Weed control in the context of integrated pest management. In: Weed Management for Developing Countries (FAO Production and Plant Protection Study - 120) R. Labrada, J.C. Caseley and C. Parker. FAO Rome.

PASSIOURA, J.B. 1973. Sense and nonsense in crop simulation. J. Austr. Agric. Sci. 39 :181-183.

PASSIOURA, J.B. 1996. Simulation models : Science, Snake oil, Education or Engeneering ? Agron. Jour. 88 (5): 690-694

PAVLISTA, A. 2002. Nebraska's Potato Industry 1985-2000. Potato eyes Vol. 14, Issue 3,

Autumn 2002.

PEDIGO L. P. and MARLIN RICE. 2008. Entomology and Pest Management, Prentice Hall. 816 p.

PENNING de VRIES, F. W. T., A.H. M. BRUSTING AND H.H. van LAAR. 1974. Products, requirements and efficiency of biosynthesis: A quantitative approach. J. Theor, Biol. 45 : 330 - 377.

PENNSTATE 2008-2009. Tree Fruit Production Guide. 248 p. College of Agricultural Sciences, AGRS -045. http://tfpg.cas.psu.edu/

PING HE , LI SHUTIAN, AND JIN JIYUN . 2009. Nutrient uptake and utilization of nutrient management for wheat and maize rotation system in North Central China. Proceeding ofthe international plant nutrition colloquium. UCD. Davis

PINTO, A. , ENGLISH H., ALVAREZ M.. 1994. Principal diseases of deciduous fruit trees in Chile. 311p. INIA. Santioago Chile.

POLLARD, KATHRINE S. 2009. What makes us human? Scientific an American. May 2009.

PRICE T. DOUGLAS. 2009. Ancient farming in eastern North America. PNAS 106 (16): 6427-6428.

RAJARAM S. and H.-J. BRAUN 2006. Wheat Yield Potential. Wheat Yield Potential Conference - CIMMYT. Obregon. Obregon. Mexico.

RAUN, W.R., and G.V. JOHNSON. 1999. Improving nitrogen use efficiency for cereal production. Agronomy Journal 91: 357-363.

REDINBAUGH M. G. and W.H. CAMPBELL. 1981. Purification and characterization of NAD(P)H:nitrate reductase and NADH:nitrate reductase from corn Roots. Plant Physiol. 68:115-120.

REGANOLD, J. P. and DAVID R. HUGGINS. 2008. No-Till: How Farmers Are Saving the Soil by Parking Their Plows. Scientific American on line, June 2008.

REVEDINA ANNA, ARANGURENB B., BECATTINIA R., LONGOC L., MARCONID E., MARIOTTI M., SKAKUNF N., SINITSYNF A., SPIRIDONOVAG E., and JI_RI SVOBODAH. 2010. Thirty thousand-year-old evidence of plant food processing. PNAS 107(44):18815-18819.

RITCHIE, J.T, GODWIN, D.C. and OTTER-NACKE, S. 1985. CERES-wheat. A simulation model of weath growth and development. Texas AM Univ. Press, College Station.

RODRIGUEZ, J. 1990. Crop fertilisation: a rational method. 406 pp

RODRIGUEZ, B. C, SEVILLANO F. G. AND P. SUBRAMANIAM. 1984. Atmospheric nitrogen fixation. A biotechnology in agricultural production. Institute of Natural Resources y Agrobiology. 65 p.

RHOADES, J.D., CHANDUVI, F. AND S. LESCH. 1999. Soil salinity assessment. FAO irrigation and drainage paper 57. Rome. 131 p.

ROSE, C.W. Agricultural Physics. Pergamon Press. London. 230 p.

ROY , R.N. A. FINCK, G.J. BLAIR AND H.L.S. TANDON. 2006. Plant Nutrition for food security: A guide for integrated nutrient management. FAO Fertilizer and plant nutrition bulletin 16. Rome.

RUSH, CHARLES M., H.-Y. LIU, R. T. LEWELLEN, R. ACOSTA-LEAL. 2006. The Continuing Saga of Rhizomania of Sugar Beets in the United States. Plant Disease 90 (1): 4-15

SAEKI, T. I960. Interrelationship between leaf amount, leaf distribution and total photosynthesis in plant communities. Bot. Mag. Tokyo 73, 55-63

SALISBURY, F. B. 1991. Lunar Farming: Achieving Maximum Yield for the Exploration ofSpace. HORTSCIENCE, Vol. 26(7): 827-833, July 1991.

SCHULTZ, T. R. and S.G. BRADY. 2008. Major evolutionary transitions in ant agriculture. Proceedings ofthe National Academy ofScience. 105(14): 5435-5440

SEXTON, P., BOME M.G., BAFLIS R.R, KAROW R., MARX E., and TOM SHIBLEY. 1999. Nitrogen rate and timing effect on irrigated Hard red spring wheat: 1999 on-farm trials in central Oregon, http://oregonstate.edu/dept/coarc/nrt.htm.

SINCLAIR, T. R. 1994. Limits to Crop Yield? Pages 1 509-532 in K. J. Boote, J. M. Bennett, T. R.Sinclair and G. M. Paulsen (editors) Physiology and Determination of Crop Yield. Crop Science Society of America, Madison, WI, USA.

SINCALIR, T. and DE WIT, C.T. 1975. Photosynthate and nitrogen requirements for seed production by various crops. Science, 189: 565-567.

SINCLAIR, T.R and N,G. SELIGMAN, 1996. Crop modeling from infancy to maturity. Agr. Jour. 88: 698-704.

SINHG, U.; BRINK, J.E.; THORNTON, P.K. AND CHRISTIANSON, C.B. 1993. Linking models with geographic information system to assist desicionmaking: a prototype for the indian semiarid tropics. IFDC Paper Series IFDC-P-19 39 pp.

STEDUTO, P., HSIAO T.C., RAES D. and E. FERERES. 2009. AquaCrop- The FAO crop model to simulate yield Response to water. I. Concepts and Undelying Principles. Agron . Jour. 101: 426-437.

STEWART, J.I., HAGAN, R.M., 1973. Functions to predict effects of crop water deficits. J. Irr. Drain. Div. ASCE 99,421-439.

STEWART, J.I., DANIELSON, R,E., HANKS, RT, JACKSON, E.B., HAGAN, R.M., PRUITT, W.O., FRANKLIN, W.T., RILEY, J.P., 1977. Optimizing crop production through control of water and salinity levels in the soil. Utah Water Research Lab. PR. 151-1, Logan, Utah, 191 pp.

STEWART SCOTT, RUSS PATRICK, PROFESSOR, ANGELA MCCLURE.2009. Insect Control Recommendations for Field Crops. 44 p. University of Tennessee. http://www.utextension.utk.edu/fieldcrops/cotton/cotton_insects/pubs/PB1768.pdf

STEWART G. R. AND T. O. OREBAMJO. 1979. Some Unusual Characteristics of Nitrate

Reduction in Erythrina senegalensis DC. New Phytologist, 83 (2): 311-319.

TAIZ, LINCOLN y ZEIGER, EDUARDO 1998. Plant Physiology. 792 p. Sinauer Associates, Inc. Publishers. Sunderland, Massachusetts.

THIRTLE, C.G. 1985. Technological Change and the Productivity Slowdown in Field Crops: United States, 1939-78, Southern Journal of Agricultural Economics, 17 (Dec.) pp. 33-42.

THOM, A.S. 1975. Momentum, Mass and heat exchange of plant communities, pp 57 -109. In: Monteith.J.L. (ed.) Vegetation and the Atmosphere. Academic Press 278 p.

THORNLEY, J.H.M AND JOHNSON, I.R. 1990. Plant and crop modelling : a mathematical approach to plant and crop physiolgy. Oxford Sci. Publ. Clarendon Press, Oxford.

TOURNAIRE-ROUX,C.; M. SUTKA; H. JAVOT, E. GOUT, P. GERBEAU, DT LUU ; R. BLIGNY AND C. MAUREL.2003. Cytosolic pH regulates root water transport during anoxic stress through gating ofaquaporins. Nature 425, 25 September, pp 393-397.

TRIPLETT, Jr. G. B. and A. D. WARREN. 2008. No-Tillage Crop Production: A

Revolution in Agriculture! Agron. J. 100:S-153-S-165.

UNDERDAHL, J., M. MERGOUM, J.K. RANSOM AND B.G. SCHATZ. 2008. Agronomic traits improvement and associations in hard red spring wheat cultivars released in North Dakota from 1968 to 2006. Crop Sci. 48:158-166.

SANTA MARIA TECHNICAL UNIVERSITY. 2007. Evaluation del Potential Productive de Biocombustibles en Chile con Cultivos Agricolas Tradicionales. Centro Avanzado de Gestion, Innovation and Technology for Agriculture. March, 2007. 148 p.

http://www.odepa.gob.cl/odepaweb/servicios-informacion/publica/Potencial_productivo

biofuels_in_Chile.pdf

USDA. 2000. Report on the Status of Medicago Germplasm in the United States - 0ctober2000. National PlantGermplasm System.
http://www.ars-grin.gov/npgs/cgc_reports/alfalfa/alfalfacgc2000.htm

VAN GINKEL, M.,I. ORTIZ-MONASTERIO, R. TRETHOWAN, AND E. HERNANDEZ. 2001. Selecting for improved N use efficiency in bread wheat.
http://www.cimmyt.org/Research/Wheat/Symp_Kronstad/posters/poster18_GinkMonas /Select_improved.htm

VAN KEULEN, H. 1975. Simulation of water use and herbage growth in arid regions. Centreforagricultural publishing and documentation. Wageningen. 176 p.

VAN KRAALINGEN D.W.G. 1995. The FSE system for crop simulation: version 2.1. Quantitative Approaches in Systems Analysis Report 1. Wageningen (Netherlands): C.T. de Wit Graduate School for Production Ecology and AB-DLO. 58 p.

VIETS, F.G. 1962. Fertilizers and the efficient use ofwater. Adv. Agron. 14:223-264

VILLAGRAN, N. y NOVOA, RAFAEL 2002. Estimation of nitrogen levels in corn by image analysis. Tierra Adentro Chile INIA. 46:12-15.

VILLAGRAN, R. 2003. Exploration of the use of high resolution images for the diagnosis of agronomic problems in grapevines. Thesis. Universidad Iberoamericana de Ciencia y Tecnologia. Santiago.

VILLASMIL P. JOSE J. 1973. Economic and Statistical Concepts in Fertilizer Experimentation. Revista de la Facultad de Agronomia (LUZ) 2(3):41-62.

VOORTMAN R.L. AND J. BROUWER. 2001. An empirical analysis of the simultaneous effects of nitrogen, phosphorus and potassium in millet production on spatially variable fields in sw niger Staff Working Paper WP - 01-04 August. 46 p. http://www.sow.vu.nl/pdf/wp01-04.PDF

WAGGONER, J.P. 1975. Micrometeorological models. 205 - 228. In: Monteith,J.L. (ed.) Vegetation and the Atmosphere. Academic Press 278 p.

WALLACH, DANIEL. 2011. Crop Model Calibration: A Statistical Perspective. Agr. Journal. 103 (4):1144-1151

WEN-GE WU, HONG-CHENG ZHANG, YING-FEI QIAN, YE CHENG, GUI-CHENG WU, CHAO-QUN ZHAI AND QI-GEN DAI. 2008. Analysis on Dry Matter Production Characteristics of Super Hybrid Rice. Rice Science 15(2): 110-118.

WILCOX CHRISTIE. 2012. Are lower pesticide residues a good reason to buy Organic?. Probably not. Scientific American. September 24, 2012. http://blogs.scientificamerican.com/science-sushi/2012/09/24/pesticides-food-fears/

WILCZEK, FRANK. 2008. The lightness of being. Basic Books, New York. 270 p.

WILLERS. 2004. GOSSYM: The Story Behind the Model. pestdata.ncsu.edu/cottonpickin/models/GOSSYM.pdf

WHITMARSH JOHN AND GOVINDJEE. 2000. The Photosynthetic ProcessIn________: "Concepts in Photobiology: Photosynthesis and Photomorphogenesis", Edited by GS SINGHAL, G RENGER, SK SOPORY, K-D IRRGANG AND GOVINDJEE, 1999. Concepts in Photobiology: Photosynthesis and Photomorphogenesis. Narosa Publishers/New Delhi; and Kluwer Academic/Dordrecht, 994 p.

WRAY, J. L., AND R. J. FIDO. 1990. Nitrate reductase and nitrite reductase, p. 241256. In P. J. Lea [ed.], Enzymes of primary metabolism. Academic.

YANG ZHENGYU, GAIL WILKERSON, BUOL, G.S., BOWMAN, D.T. and R.W. HEINIGER- 2009. Estimating Genetic Coefficients for the CSM-CERES-Maize Model in North Carolina Environments. Agron J. 101:1276-1285.

YEATES, NTM. 1955. Photoperiodicity in cattle. I. Seasonal changes in coat character and their importance in heat regulation. Australian Journal of Agricultural Research 6(6) 891 - 902.

YEATES, NTM. 1957 Photoperiodicity in cattle. II. The equatorial light environment and its effect on the coat of European cattle *Australian Journal of Agricultural Research* 8(6) 733 - 739

ZAGAL, ERICK V. 2005. The nitrogen cycle in the soil. Revista Ciencia Ahora No. 16: 103-116. http://www.ciencia-ahora.cl/Revista16/14ElCicloDelNitrogeno.pdf.

ZEDER, A., MELINDA. 2008. Domestication and early agriculture in the Mediterranean Basin: Origins, diffusion, and impact. PNAS 105 (33): 11597-11604.

MIX
Papier aus verantwortungsvollen Quellen
Paper from responsible sources
FSC® C105338

Printed by Books on Demand GmbH, Norderstedt / Germany